MW01132840

# Emergency Evacuation Planning for Your Workplace:

## From Chaos to Life-Saving Solutions

**Jim Burtles**, KLJ, MMLJ, FBCI

**Kristen Noakes-Fry**, Editor

ISBN 978-1-931332-56-9 (Hardcover)
ISBN 978-1-931332-67-5 (e-Book)

a division of Rothstein Associates Inc

Brookfield, Connecticut USA
www.rothstein.com

*Paid purchasers of this book are entitled to a free download of extensive supplemental licensed materials upon registration. See instructions on back page.*

ISBN 978-1-931332-56-9 (Hardcover)

ISBN 978-1-931332-67-5 (e-Book)

Library of Congress Control Number
(LCCN) 2013933852

a division of Rothstein Associates Inc

Philip Jan Rothstein, FBCI, Publisher
4 Arapaho Road
Brookfield, Connecticut 06804-3104 USA
203.740.7444 • 203.740.7401 fax
info@rothstein.com
www.rothstein.com

*Paid purchasers of this book are entitled to a free download of extensive supplemental licensed materials upon registration. See instructions on back page.*

# Acknowledgments

During the creation of this book, a number of people have supported and contributed directly or indirectly to its contents. Without their intellectual generosity, this book might never have been completed, or it would have been a much smaller work. While some have been close friends and colleagues, many of those who came forward with ideas, stories, or examples were unknown to me and simply wanted to contribute to the body of knowledge. I wish to thank them on behalf of the readers and all those who may eventually benefit from their wisdom.

I set out from the beginning to offer sound practical advice based on experience, not seeking to impress nor to become too philosophical or academic; the aim has been to help the reader prepare for, and deal with, emergency situations where there is a need to evacuate volumes of people.

## Special thanks go to:

My wife, who has been a constant supporter, source of encouragement, and willing listener throughout this whole project.

My publisher, Phil Rothstein, and my editor, Kristen Noakes-Fry, who guided and directed the development of the final version of the text. Kristen in particular was extremely helpful, diligent, and patient throughout the development of the final set of materials. The resultant high quality of this work is largely due to her support and encouragement.

My good friend Peter Jack, for acting as a sounding board and a source of fresh ideas. He was also instrumental in making the material on the Freemasons' Hall available for publication.

Lynda McMullan of Lincoln University, who was very helpful with regard to enabling us to publish information about her experience and achievements on behalf of the university.

Mark L Friedman, MD, who was kind enough to grant permission for us to quote from his book *Everyday Crisis Management* (2001).

Others whom I must thank for their contributions include Mel Gosling, John Glenn, Colin Ive, Jace Mudali, and Robin Gaddum.

# Dedication

This book and its contents are dedicated to all those companies and municipalities across the world who have in place the policies and procedures which are designed to ensure the safety and welfare of their employees, residents, and visitors, together with all those emergency professionals and local volunteers who regularly evacuate thousands of people from situations of danger and help them to reach places of safety during the disastrous events of so many kinds and sizes that seem to occur wherever people gather throughout the world.

It is also dedicated to the people of New York City and the surrounding areas, who stayed strong and resilient in the aftermath of the World Trade Center attack in 2001 and who remain tough and resourceful as they rebuild and recover from the widespread damage of Hurricane Sandy in 2012.

# Preface

As this book was being readied to go to press, I was shocked and angered to see a *New York Times* headline, "More Than 300 Killed in Pakistani Factory Fires." Having spent the last decade or so of my career as an advocate for workplace safety, I was sickened and frustrated to read that the workers inside the textile factory "had few options of escape – every exit but one had been locked...and the windows were mostly barred." Workers had flung themselves from top floors of the four-story factory. Most died from smoke inhalation.

This tragedy in Karachi, Pakistan, on September 12, 2012, now ranks as one of the worst industrial disasters in history – killing twice as many workers as the eerily similar Triangle Shirtwaist Factory Fire of March 25, 1911, in New York City, in which 146 garment workers, mostly young, immigrant women, died from fire, smoke inhalation, or falling to their deaths. They had been unable to escape because managers had locked the doors to stairwells and exits to prevent pilferage. (And since I wrote these lines, a fire in a garment factory in Bangladesh in November, 2012, killed over 100 workers, also trapped without access to emergency exits.)

Reading about the recent fire and working through my anger and frustration, I reflected that, a century later, in spite of all the efforts that many of us have made at international standards for worker safety, 300 innocent people were permitted to die because their employers – who were responsible for their health and wellbeing while they were on the job – had not implemented basic emergency evacuation procedures. Almost exactly a century ago, there was an infamous tragedy of trapped workers in New York City's garment district - and now, in 2012, it happens again in another garment district in Karachi, Pakistan – and I have to wonder why hard-working factory employees are no safer now than they were 100 years ago.

## Origins of This Book

My basic ideas for this book began, not a century ago and not last week, but over a decade ago in the aftermath of the terrorist attacks on the World Trade Center. I was already thinking seriously about emergency evacuation when a client, a major international news agency, asked my company for help. We had already been working with the agency on a business continuity program, but now they were concerned that their offices around the world could also become targets for terrorist activities. They asked for a specific methodology that would permit them to roll out an Emergency Evacuation Plan (EEP) for each of their 400 offices around the world.

Since we did not have such an EEP methodology on the shelf, we plunged into the development with some enthusiasm, and in six or seven weeks we were ready to make a firm proposal. But by then, the New York office of the agency had recovered from its fright and had moved on to look at "other more important things."

At this point, I was intrigued, and began to incorporate the ideas I had developed in that project into my regular work as an independent Business Continuity (BC) consultant. I quickly found myself becoming an advocate for EEP as a logical extension of Business Continuity Management (BCM). These fundamental principles have remained unaltered, but experience and reflection have refined and expanded the subject considerably.

During the final stages of the preparation of this book, I began to see a clear and distinct methodology emerging, a formal, auditable process for EEP. I adapted the Business Continuity Institute (BCI) lifecycle, which represents best practices in BC, to create a unique EEP lifecycle. I explain this lifecycle in the introductory chapter which follows, and I use it as the basis of the structure of this book, dividing it into six sections, reflecting each of the six phases of the lifecycle.

## Benefits of Planning Ahead

Whether the trigger is a fire, flood, explosion, earthquake, or some unforeseen event is irrelevant – at some point, you will need to get everybody safely out of the building and deal with the effects and consequences of the cause, whatever it might be.

Unfortunately, many management teams seem to consider planning for a worst-case scenario to be a waste of time, or they rank emergency preparedness very low on their list of priorities. If you take responsibility for this aspect of the operation, your challenge is to persuade the powers-that-be of the importance of these matters. As you encourage those in power to accept your point of view about emergency preparedness, I recommend the traditional carrot-and-stick approach, beginning with the "stick."

As examples, you could use statistics and potentially worrying facts about the frequency and unpredictability of disasters and their ominous consequences. Calamities such as 9/11, Hurricane Katrina, and Japanese earthquakes may be ignored as being unique and unlikely to happen again, especially not locally. However, a little research will yield many examples of emergency situations in which people had to get out quickly - fires, flooding, storm damage, explosions, pollution, structural failure, accidents, and crashes.

Management should find it thought-provoking that the cost of financial recovery from an incident is much reduced if all the preventive and protective measures are in place. In addition, the cost in work-hours will be lessened because a smooth exit reduces the scale of the backlog of work that builds up during and after an emergency situation. This information should help management to see that good planning and some training enables an organization to predict exactly what needs to be done to restore the business operation fully and to be able to use this information in an actual emergency.

An enormous goodwill benefit can be gained from the confidence of staff and visitors who appreciate that someone cares enough to consider their health and safety and provide them with a safe working space. Such confidence does a lot to enhance and cement the team spirit, while encouraging loyalty, cooperation, and trust.

Whether top management buys in to these arguments or not, the one factor they cannot ignore is fiduciary responsibility. Should the lack of an effective EEP result in loss of life, serious injury, or significant financial loss, such an outcome could represent a significant failure of fiduciary responsibility. A failure of this magnitude could leave the organization – and the individual members of the management team – open to potential civil or even criminal regulatory or legal liability, as well as costly and painful litigation.

## How to Use this Book

While this book has been written for a commercial audience, the underlying concepts apply to any type of organization, large or small, whether profit or not-for-profit. All organizations are responsible for making effective use of the cash flowing through their operations and for protecting their image and limiting damage. These concerns entail keeping people safe at all times.

While the information in this book may be applied to evacuating any kind of structure, the book does not take into account the specific regulations, recommendations, and standards of practice which apply to residential premises, intended for permanent occupancy and used for domestic purposes. In most countries, the preferred emergency evacuation procedures and preparations for residents are significantly different from those which apply to non-residential properties. Because the design and construction of residential premises is considerably different from that of other types of premises, a number of different factors affect the way in which an emergency response should be organized. If you are considering the development of emergency plans for a residential complex, I recommend that you supplement this book by consulting a fire safety engineer (who should be consulted for any facility or complex), who will be able to advise you regarding the current regulations and any relevant recommendations which might apply in your case.

## Final Thoughts

An emergency is an extreme problem or an extreme set of circumstances, both uncommon and unfamiliar. As human beings, we are never fully prepared for such an unusual event. We need information, support, and guidance. Often there seems to be nobody to turn to, while time is at a premium. At the corporate level, we are concerned about damage to property and the longer-term financial implications, however, our primary concern and responsibility is to ensure the health and safety of anyone who may become trapped in our building – to evacuate the building as quickly, safely and efficiently as possible. This substantial responsibility for human life, and welfare can weigh heavily on the unprepared; the consequences of inaction, hesitation, misunderstandings, or errors of judgment can be very serious indeed.

Throughout this book, I have used the term emergency evacuation to mean "an organized escape from danger to a place of relative safety." It is not simply the action of moving populations, but taking complete responsibility for the wellbeing of occupants in an emergency. By following the methodology in this book, you can reduce risk dramatically and improve the chances of achieving not only a successful recovery of the business operations but also the health and safety of all those concerned.

Jim Burtles
London, England
October, 2012

# Foreword

Organizations talk about the need to have effective emergency response and evacuation plans and procedures in place. However, the need to convert this talk into action cannot be more clearly emphasized than by looking at the actions at Morgan Stanley in the second tower of the World Trade Center on 9/11. The strategy, plans, and procedures that Rick Rescorla established did enable 2,700 Morgan Stanley employees to safely evacuate. Morgan Stanley obviously had put in place a process and program which provided for a successful evacuation in a desperate situation. This is a prime example of the benefits of using a similar type of process and program that Jim Burtles expounds in *Emergency Evacuation Planning for Your Workplace: From Chaos to Life-Saving Solutions.*

Many years ago, Jim Burtles and I were both part of the group that established the Business Continuity Institute (BCI) in the United Kingdom. We shared the same philosophy – that Business Continuity Management (BCM) did not just relate to maintaining business functions but really started at or even before the incident occurred, particularly if there was a warning period.

History shows that, to be effective, an organization needs a structured, formal process to be employed in the development of emergency response and evacuation plans and procedures. The follow-on question is, What exactly should that process be? Of course, a number of regulations, standards, and guidelines relate to this subject, but these usually detail what should be included or covered, rather than offering guidance in the step-by-step development process.

As Jim points out in the introduction to Phase 5 of this book, companies often fail to take an orderly approach to the process. Too often, the actual Emergency Evacuation Plan (EEP) is regarded as the starting point for the process, while the critical first four phases that Jim outlines here are ignored. Writing a plan without the preparatory research and training is tantamount to diving off the high platform, or "into the deep end," as he puts it, without any previous training.

The development process encompasses not just development, but information gathering, strategy development, awareness creation, exercising, and testing – just to mention the basics.

Typical existing explanatory material is often in the form of official handbooks or other formal documentation. What is painfully lacking or inadequate is a single source of good practical explanations, tips, examples, and customizable documentation. In this book, Burtles addresses that need. Not only does he give insights into what should be covered, but he provides a wealth of downloadable documentation that can be customized for an organization's specific needs.

I noted many important aspects of this book which are often not found elsewhere; for example,

- The use of a clear development lifecycle similar to BCI's business continuity management model. This model emphasizes the integration of BCM and EEP, applying the best practices reflected in the BCI model to researching, planning, and exercising a formal EEP.
- The reference, explanation, and use of the Plan-Do-Check-Act concept in EEP. This has already become a central point of reference for ongoing management systems and is now very much referenced in current business continuity standards.
- The real-world discussion questions at the end of each chapter which, encourage readers to apply the concepts in the chapter to their own business experience and observations, facilitating further consideration, discussion, investigation, and thought.

Additionally, topics often not addressed in detail elsewhere are covered very specifically in the book; for example,

- Consideration of regional evacuations, evacuation of downtown business areas, evacuation of college campuses and recreational complexes – not just evacuation of individual buildings. (Jim covers EEP for areas which are occupied by the same population most of the time, as well as for areas that will be occupied primarily by visitors, shoppers, and guests.)
- Potential post-evacuation issues and considerations. (Burtles makes clear the EEP responsibility for the wellbeing of people, emphasizing that it entails every step that needs to be taken from the moment the alarm sounds until all the people involved are

safely back at their desks, back in their homes, safe in an emergency shelter, or have become the responsibility of some other agency.)

- The importance of Available Safe Egress Time (ASET) versus Required Safe Egress Time (RSET). He asks readers to question the general basic assumption (made by most organizations in their planning) that there will be adequate time for evacuation. But is that really true? Use Jim's formulas to compare ASET and RSET under various scenarios and see the results. Your findings may be very interesting and hopefully will help you to think outside of the box with your evacuation plans.

This new book will be a wakeup call, not only for those involved with BCM and EEP, but also anyone involved with employee safety, emergency response, physical security, and facilities management. Ask yourself: Does your organization have the coordination between everyone as outlined in the book? Even if you do, is it to the depth that it should be? Use the guidelines in this book to assess the coordination and determine what else needs to be done. If you do not have such coordination, why not? This is the first major gap in your program. Start working on it immediately! Having worked in an organization in which the need for such coordination was recognized, I cannot overemphasize its importance and how much it helps when an emergency arises.

For those of you with existing plans and procedures, use the book as an audit tool. In developing the existing documentation, was the process structured as a program or were individual sections put together over a period of time with no real linkage or flow? Ask yourself if a process similar to Jim's was used in developing the plans and procedures. Was similar information gathered and used? Is the whole thing considered as a continuing program or a plan in a binder sitting on a shelf?

From the time I first met Jim Burtles, it was evident that he had a keen interest in EEP. Over the years, Jim has accumulated a wealth of knowledge on the subject and now, in *Emergency Evacuation Planning for Your Workplace: From Chaos to Life-Saving Solutions*, I am pleased that he has taken the opportunity to pass on his experience and expertise.

Melvyn Musson, FBCI, CDRP
Retired Senior Business Continuity Planning Manager, Edward Jones
St. Louis, Missouri, USA
March, 2013

# Foreword

We have all been involved in the occasional fire drill, which, in most cases, involves a reluctant response to the test alarm and a leisurely stroll towards a muster point, while we glance up at the sky hoping that we'll all be back in the building before it starts to rain. Unfortunately, these exercises tend to be perfunctory – all that happens is that at a pre-arranged time, given a pre-arranged signal, a group of people wanders toward the nearest fire exit and plods towards the nearest muster point (or, perhaps, some of them simply follow the people in front, assuming that they must be heading towards the muster point).

Such drills are generally regarded as simply an amusing break in the work routine. In real-life situations, when it is not a "drill," these blasé attitudes commonly displayed during the "drill" are quickly replaced by stress and confusion. In a real emergency, it is only proven frameworks and calm confidence fostered by in-depth practice that can ensure the leadership required to avoid panic and chaos.

Personally, I will never forget the 7 July 2005 London bombings (often referred to as 7/7) when terrorists struck London's public transport system. That particular day, I was working in Bishopsgate – a stone's throw from Liverpool Street mainline station, a transport "hub" connecting commuter mainline services with the London Underground and bus services. The location was also very close to Aldgate tube station, where one of the suicide bombers detonated an explosive device. In all, 52 civilians and the 4 bombers were killed in the attacks, and over 700 more people were injured.

By pure chance, I was close enough to observe but far enough away to be out of immediate danger. One of the everlasting impressions that will remain with me from that time was the concern and confusion so evident on the faces of the people I observed – and not just those who were directly affected. One thing that was made clear during the subsequent investigation and inquiries was how well the evacuation process was handled and, in particular, the effective actions that were taken to avoid widespread panic. The crowd control at the various incident scenes together with the efficient evacuation of those immediately affected by the incidents was handled by emergency services staff in cooperation with London Transport staff.

Essential to the success of the emergency response on 7/7 was familiarity with the procedure. The UK emergency services are, of course, well trained to deal with crowds of people during a major incident; likewise, public transport staff members receive training to ensure safety during an evacuation of public areas. On the other hand, for many others who undertake the responsibility to act as marshals in the event of a building evacuation, the training is often less rigorous and less frequent, leaving them less well equipped to ensure that workplace evacuations are performed effectively.

This gap in knowledge and training is exactly why Jim Burtles' book, *Emergency Evacuation Planning for the Workplace: From Chaos to Life-Saving Solutions*, is so important and long-overdue. The book provides a carefully wrought framework to formalize an activity that needs to be both structured and managed.

Jim's book meets a most critical need by providing a well thought out and structured approach to emergency evacuations. With his long experience of business continuity planning, Jim has applied his in-depth knowledge of the evacuation process, providing a detailed, yet digestible, approach to emergency evacuations. In this book, you will learn the key attributes of a robust and effective approach to building evacuation, what it takes for people of all abilities and limitations to escape from buildings of all sizes under a variety of conditions. He lays out a six-phase methodology for developing, testing, and maintaining an effective evacuation plan.

Jim's real-world experience also enables him to provide insight into the best practices to ensure that, once it is defined and implemented, the plan is sustained via effective management. He covers policy development together with processes for ongoing assurance and feedback. This approach creates that all-important management system to ensure that your procedures remain effective and that the investment you made in creating them will be sustained.

The process of emergency evacuation has been overlooked for far too long. We cannot expect to be able to evacuate our workplaces effectively without a commitment to devoting time and resources to develop appropriate processes to support it. Yet, most traditional business continuity and incident management plans tend to assume that an effective evacuation will take place somehow, even when the plans have given very little attention to the process of actually getting staff and visitors out of the building to a place of safety so that the all-important roll-call can be performed.

I am still haunted by memories of 7/7 and what might have been my fate if I had been only a few blocks closer to the explosion – and what would have been the fate of thousands of London commuters if trained emergency workers had not been present to handle the evacuation. These memories make me grateful for the framework that Jim has provided in *Emergency Evacuation Planning for the Workplace: From Chaos to Life-Saving Solutions*. He offers a reliable approach to ensure the safety and protection of our people in the workplace - who are, after all, our greatest asset in a major incident.

Steve Dance, CISA
Owner, RiskCentric;
Designer of compliance, risk management frameworks, and management systems.
Chair, Business Continuity and Security SIG, British Institute of Facilities Managers (BIFM)
Peterborough, United Kingdom
February, 2013

# Foreword

Ten years later, the image still haunts me – I am standing atop the rubble of the World Trade Center, September 2001, and I look up at a building that was adjacent to the towers, and see through what is left of a wall. Shattered windows let me see intact file cabinets and office furniture covered with the grey dust of catastrophe, now open to the elements. But what attaches to my mind forever is a sweater that is hanging from the back of a desk chair. Someone was in that chair. She was working at that desk. Where is she now? Who was she? Did she evacuate? Was she instructed to exit quickly and quietly at the first sign of trouble? Or did she miss her bus that morning? Did she ever go back and get that sweater? I'll never know. But what sticks with me will always be the question of how or if she had been informed, or if she was forced to improvise her escape from hell?

Evacuation strategies can range from simple to complex. A very long time ago I was eating Chinese food and opened my fortune cookie to read, "When you enter, choose your exit." I have used this phrase as a simple personal evacuation and exit strategy since. A more complicated need exists when trying to move large numbers of diverse people to safety during an emergency. My simple practice is useful because it reminds me to stay awake and alert to possibilities and surprises. Whereas a personal, simple, fortune-cookie exit strategy is useful, a carefully planned evacuation system is the culmination of critical thinking in the face of an emergency event. Such strategy is especially necessary for moving and managing large groups of people quickly.

In this book, Jim Burtles offers a well-considered, clearly-crafted technical imperative that considers a broad range of emergency evacuation. As an expert on the emotions of humans before, during, and after disasters, I found myself asking my most important "how-to" professional questions of this book, and within a few pages, I was able to find a more than appropriate detailed approach to address my concerns. Although Burtles is not a mental health professional, he has been driven by his own experiences and human compassion to find and report a comprehensive methodology to protect humans from harm in the case of an emergent need for evacuation. His text brilliantly does what I call "Fire Extinguisher Thinking." Most people have a fire extinguisher on hand so that they don't need to spend their days paranoid about fires. They are ready to the best level they can be. They have prepared and can move forward. Burtles has created a substantial and tight infrastructure for planning, practice, and procedures that can direct any company, no matter the size, to evacuate from both the imaginable and the unimaginable events. This exemplifies my concept of Emotional Continuity Management© at its best: good for business and for humans.

As a trauma counselor who has worked at the sites of major disasters, I attempt to communicate through my books, classes, and seminars the messages of the dauntingly unpredictable nature of human behavior during a crisis. Nice people do not want to think about awful horrible things. I respect that. I don't either. But I don't have the luxury of denial. I get cold chills when I recall helping people heal from the bizarre things they have seen or done during crisis and disaster. Many of their stories could have been prevented with even the simplest of forethought and planning. I know it is difficult to imagine yourself or other sane and nice colleagues becoming completely deranged during a crisis. It is equally difficult to consider the weirdest, most annoyingly dysfunctional, coworker becoming the heroic leader during an emergency. But it happens. Sudden emergencies can trigger extreme altered mental states that can lead to completely unpredictable behaviors. This is why I believe that excellent planning is not optional. Today, planning is no longer a luxury.

When in an altered state, most people do not think clearly about options and are overwhelmed by their sudden powerlessness. They can do the most stunningly bizarre things to try to get back into control. I have stories! Some stories are heroic, sweet, and charming, and some I will never share because they are too heinous.

In this book, Burtles provides a strong framework for the practice, and (practice and practice and practice and practice) of well-planned evacuation drills and procedures, the writing and re-writing of strong policies, buy-on from top to bottom and from bottom to top, good signage, and clearly understood and agreed-upon methods that can give people at risk a sense of personal power that can significantly increase their chance of survival. People with options tend to make different and better decisions than victims who feel powerless. If something unexpected happens and someone remembers the plan, others will also. Someone does the smart thing because a well-constructed, tediously drilled plan is already in place, based on as many possible scenarios as humanly imaginable. Solid! Others will snap back to clear thinking, take the prepared action, and support others in shock and confusion to apply the plan. This clarity provides the best hope that the majority of people will follow to safety. These well-rehearsed behaviors become the default, the normal action in the middle of massive abnormality. There won't be the question, "What are we supposed to do?"

Sometimes there just isn't enough time to ask the question. Seconds can count. Emergency responders know this all too well. They drill endlessly to shave microseconds off of their response times because they know that an excellent response increases a sense of personal empowerment smack dab in the middle of extreme powerlessness and enhances the chances of survival exponentially. And clear thinking can also lead to emergent improvisational behavior that may be required in the presence of unanticipated events. Victims have no choices. Survivors have choices and take them.

Jim Burtles gives us a book that is a comprehensive, didactic, technical, organized, blueprint for best chances for physical survival during any event that demands evacuation. It offers best practices for getting out of harm's way. I share and appreciate Jim's deep concern for the well-being of people at risk. I know that the long-term emotional issues of survivors can be managed later by professionals in the healthcare and therapeutic industries. But first, we need to have the people out of the building! First things first! Burtles' book is a thorough recipe for the planning, design, implementation, movement, and management of people from harm's way to safety so that they can have a story to tell.

Once upon a time, when I was responding to a disaster, the emergency management professional providing the briefing suggested that if any of us saw a large group of people running, we should start running with them in whatever direction they were going because they probably knew something we didn't. This seemed a rather interesting, inspired, and improvisational evacuation strategy. I value improvisation. I also value planning. I suggest that you take Burtles' book to heart and embed a system that will give you a creative vision for survival for the people you hope to protect. People will be people, and disasters have a life of their own. Indeed, there is never any way to fully predict, or protect people from, the variables of emergencies. However, that being said, if you take the time to read, study, contemplate, and apply the tactics from this thoughtful and elaborate text, you may well become your company's hero. It takes energy and stamina to stand up for human safety and survival, physical and emotional, and you can take this book as a strong ally into the fray, because it is a cutting-edge methodology to plan and prepare for emergency evacuations.

Exiting and entering are powerful experiences of human life that most people just take for granted. As a mental health practitioner, I find that many of the issues that bring people into counseling are about difficulties during life's transitions. I am blessed with being able to midwife a slight nudge forward away from their pain, crisis, and confusion, and to stand by in the sidelines to cheer as they rediscover the rest of their wonderful lives. We all get stuck and need, as Burtles puts it, "good signage" from time to time. This book is good signage and more than a nudge toward good thinking. This book is a parachute. I hope the person at the desk in my haunting building image got a nudge to get the hell out – and I hope that she is wallowing in gratitude while living a wonderful life somewhere delicious without her sweater.

Vali Hawkins Mitchell, PhD, LMHC
Seattle, Washington, USA
September, 2013

*Dr. Hawkins Mitchell is a Certified Traumatologist, Licensed Mental Health Counselor, Consultant, and Executive Coach. She travels widely, providing individual and group trainings as the leading authority in the field of Emotional Continuity Management and Emotional Terrorism in the workplace. Author of the book The Cost of Emotions in the Workplace: The Bottom Line Value of Emotional Continuity Management (Rothstein Associates, 2013). She is a frequently published writer and appears regularly as a public speaker.*

# Contents

**PHASE 4 - Determining Evacuation Strategy** 95

# Introduction

## The Essentials of Emergency Evacuation Planning

Emergency evacuations do not always have happy endings. Front-page headlines regularly feature mismanaged evacuations:

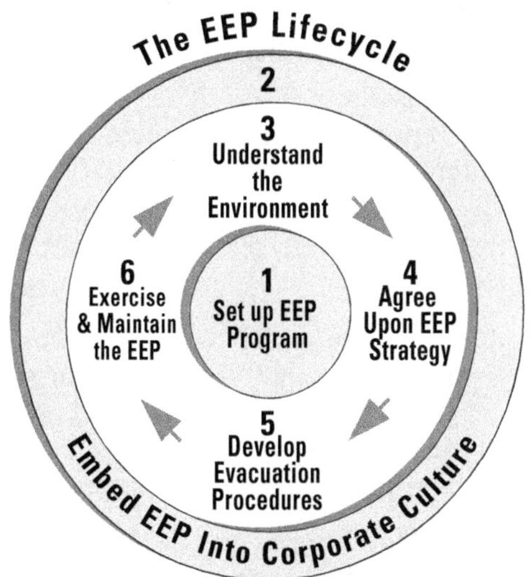


- Investigations continue into the 25 deaths and disorganized evacuation attempts that followed the January, 2012, accident involving the Italian cruise ship Costa Concordia. Negative publicity over the following months hurt the cruise ship company and the entire cruise industry.
- In the wake of the April, 2007, Virginia Tech tragedy in which a gunman killed 33 university students and faculty, public criticism was directed at university administration because of a lack of timely notification and a delay in evacuating students. Investigations, fines, and lawsuits continue. In March, 2012, a jury awarded families of two slain students $4 million each in a wrongful death lawsuit against the university.

# 0.1 Are You Prepared?

Whether your role in the organization is:

- Director,
- Owner,
- Partner,
- Legal Counsel,
- Facilities Manager,
- Business Continuity Manager,
- Emergency Response Manager,
- Human Resources Manager, or
- Leader of a Workgroup or Department,

...you need to take a minute to answer these questions to evaluate the status of emergency evacuation planning (EEP) in your organization.

## 0.1.1 Does Your Company Have Effective Emergency Evacuation Plans and Procedures in Place?

Have you seen reports of regular emergency evacuation exercises or drills? Those reports would have included recommendations for the improvement, maintenance, or continuation of the plans and procedures.

**Without a formal, signed, published policy document which quotes a benchmark or standard...you and your colleagues could be accused of corporate negligence.**

*Is the answer "Yes"?* – If your organization has formal, published plans and procedures in place, take the time to review these in accordance with the lifecycle outline in this book to bring them in line with current best practices.

*Is the answer "No"?* – You have a problem! If those plans and procedures are not in place, an investigator (or auditor) should become familiar with the concepts outlined in this book before conducting an investigation or audit of whatever emergency procedures might be in place.

## 0.1.2 Does Your Organization Have a Formally Agreed Upon Policy Regarding Emergency Evacuation of Your Premises?

Without a formal, signed, published policy document which quotes a benchmark or standard designed to be used for reference purposes, you and your colleagues could be accused of corporate negligence.

If there is a policy document in place but it makes no reference to a benchmark or standard, then you should appoint someone to investigate the possibility of incorporating this EEP lifecycle model into your policy. The lifecycle model presented here is a useful representation of current best practices.

*Is the answer "No?"* – You have a problem! If such a document does not exist, then you need to develop a policy without delay. Appoint someone to apply the concepts of this EEP Toolkit lifecycle model and prepare a draft policy which you and your colleagues can either approve or amend. You can find a suitable template in the accompanying toolkit which can be adapted to suit the needs of your organization.

*Is the answer "Yes, but"?* – If there is a policy document in place BUT there is no reference to a benchmark or standard then you should appoint someone to investigate the possibility of incorporating our EEP life cycle model into your policy. Our life cycle model does represent the latest and best practice in this regard.

### 0.1.3 Could Your Company be Deemed Guilty of Negligence in Regard to Protecting the Health and Safety of Those Who Use, Visit, or Reside in Your Premises?

This question is especially important in regard to the disabled, who are particularly vulnerable and are normally the subject of specific laws and regulations. You should have adequate, i.e., tried and tested, emergency evacuation plans in place to protect the vulnerable. Look for hard evidence on file that verifies specifically that these plans have been tested and kept up to date.

*Is the answer "No"?* – If some of these measures are already in place, then your appointee should evaluate and improve them in accordance with the practices covered in this book.

*Is the answer "Yes"?* – You have a problem! If such a document does not exist, then you need to develop a policy without delay.

In evaluating your level of preparedness and what needs to be done next, the EEP lifecycle model summarized below – and covered in detail in the six sections of this book – should be taken as your guide to the latest and best practice in the subject.

## Ten Signs of an Effective EEP Program: A Checklist

1. Single point of responsibility – Someone is responsible and accountable for the safe evacuation of everybody who may be on the premises.
2. Trained marshals in all areas – People can rely upon well trained helpers to guide them and assist them to leave the premises in safety.
3. Well marked exit and escape routes – All of the exit routes, exits and escape routes are clearly marked and indicated to suit the needs of the population.
4. Multiple safe assembly areas – A number of alternative safe assembly areas are easily accessible for everybody who may be on the premises.
5. Protected exit points – All exit points are protected by a stout canopy to protect evacuees from falling masonry and other debris as they leave the premises.
6. Visitor awareness program – All visitors to the premises are properly informed about evacuation procedures and planned drills or exercises.
7. Published regular exercise and testing program – A regular schedule of exercises, tests, and drills is in place, and everyone is kept informed about it.
8. Effective personnel accounting procedure – Tried and tested procedures are in place to ensure that everybody is properly accounted for in an emergency.
9. Ensures that all people on the premises are aware of an emergency situation and how they should respond.
10. Post-incident support in place – Support and assistance will be available to all those who may have been affected mentally, physically, or spiritually.

## 0.2 Summary of How the Emergency Evacuation Planning Lifecycle Works

Because this is a large, important, and lengthy project or program, I recommend you start with a game plan to put everything into perspective long before you get into such detail that it could be easy to lose sight of your original objectives. What follows is a high-level model which illustrates the main concepts of what we are trying to achieve.

The emergency evacuation planning (EEP) lifecycle as shown here is derived from the original business continuity management (BCM) model for creating a business continuity plan (BCP), which has been adopted by both the British Standards Institution (BSI) and the Business Continuity Institute (BCI) in both the UK and the US. It is used as the basic model in the British Standard BS 25999, published by the BSI, and it also forms the basis of the Good Practice Guidelines (GPG) published by the BCI. If EEP is considered a subset of BCM, then it follows that this derived model should be a useful EEP guide. Even if we think of EEP and BCM as separate but parallel activities, this model is worth considering simply because it has been shown to work so very well in actual practice.

**The step-by-step...approach to EEP...is very similar to the BCP process. The key difference is that EEP is primarily concerned with people, warmth, and shelter rather than documents, data, and operations.**

Because the process is an ongoing one, the EEP lifecycle is portrayed as a circle to suggest continuity or a never-ending loop. However, I prefer to think of it as a rolling wheel which needs a regular nudge to keep it going. The step-by-step, organized approach to EEP outlined in this and the following sections of this book is very similar to the BCP process. The key difference is that EEP is primarily concerned with people, warmth, and shelter rather than documents, data, and operations. Each of the following six sections of this book will explore one of these phases in depth.

### 0.2.1 Phase 1 – Set up the Emergency Evacuation Planning Program

We start off in the center or at the hub of the lifecycle, marked "1" in the diagram. This initial phase is key. Without the support of proper authority and budget, your progress will be rather difficult and success most unlikely. Before anything constructive can begin to happen, you need to get permission to go ahead with an EEP program, and that requires a senior decision and endorsement of a policy to guide the rest of the work. Depending on the culture and hierarchy, as emergency manager, you will need a sponsor or a champion whose authority and respect will ensure the smooth running of the program and cooperation of others. Although the majority of the actual planning work may be expected to fall on one pair of shoulders, you need to make sure that responsibilities for various aspects of emergency planning and management are formally assigned and properly accepted by those involved.

## 0.2.2 Phase 2 – Embed EEP Into an Aware and Prepared Corporate Culture

The second phase (marked "2" in the diagram) forms the periphery of the lifecycle diagram. This shape and position is meant to suggest that this is the ongoing process which binds the whole thing together and keeps it alive. Before any plan can be effective, people have to become aware of it and have faith in its validity. This awareness campaign should begin as soon as there is a policy in place. You should use any and all means of communication available to get the message out into the community which you are serving with this program.

## 0.2.3 Phase 3 – Explore, Assess, and Understand the Environment

Phase 3 (marked "3" in the diagram) is one of the four core activities in the ongoing evolution of the EEPs and procedures, matching them to the ever-changing needs of the community you are protecting. Although those changes may not always be spectacular or obvious to the outsider, you do have to stay in touch and move with the times to provide the best level of cover and support in the event of a serious incident. Here the really serious work begins. You need to explore which functions of the organization require emergency response or emergency management plans and what areas need to be considered within your EEPs. If you are dealing with a large complex of buildings or an especially large building, you will need to prioritize the sequence of developing the detailed strategy for each of those areas which come within the scope of your program, as outlined earlier in the policy during Phase 1.

## 0.2.4 Phase 4 – Agree Upon an Evacuation Strategy

Phase 4 takes you through the development of the strategy, introduces you to the tools and techniques, and acts as your guide and mentor throughout the process. This phase 4 (marked "4" in the diagram) is on the right side of the lifecycle, is the point at which you start to focus on developing the evacuation strategy for your organization. It is the start of the in-depth EEP, which begins with a visit to each of the potential exit points to determine if each one is safe enough to be used in an emergency when large volumes of people will want to get out in a hurry. In effect, this stage is a thorough survey of premises and the surrounding environment to look for potential threats which could trigger an emergency and for risks which might compromise the evacuation of the premises. Therefore, it is worth considering working with those groups who are familiar with the place and its people, e.g., facilities management, maintenance, security and HR. Their first-hand knowledge and experience will prove invaluable as you identify the potential problems and explore the possible solutions.

**The main question is whether a single generic plan will suit all needs or whether there is a need to consider a suite of plans to cover different parts of the establishment or perhaps the variety and type of people and their possible whereabouts.**

## 0.2.5 Phase 5 – Develop Evacuation Plans and Procedures

In Phase 5 (marked "5" in the diagram) as depicted at the bottom of the lifecycle model you are approaching what many would regard as the principal activity of this whole program. It is certainly the phase where you will be developing and delivering the most tangible products of the program. The main question is whether a single generic plan will suit all needs or whether there is a need to consider a suite of plans to cover different parts of the establishment or perhaps the variety and type of people and their possible whereabouts. Some situations may warrant individualized solutions such as PEEPs, or personalized plans. While this type of thinking should have been initiated at the policy development stage, you may find yourself modifying your original thoughts as you gain a better understanding of the type and scale of the problems involved.

## 0.2.6 Phase 6 – The Ongoing Program: Exercise and Maintain the EEP

In Phase 6 (marked "6" in the diagram) you are entering the long-term phase of the program. While we are describing it after the other phases, you must avoid thinking of it as the last or final phase because it is the ongoing process which keeps the whole survival-based program alive. Without regular exercising, reviewing, and updating, your plan faces the distinct danger of an evacuation failure which could have disastrous consequences. Regular exercises should then be run to familiarize everybody with the plans and choices. Normally, these would be run at least once a year, preferably twice – typically once in summer and once in winter conditions. You may find that you need to run them more frequently depending on the volatility of the population and the capability of individuals to retain such information. Where there is a disabled sector of the population, you may find it necessary to run limited exercises for them and their helpers; it might not be necessary to involve the rest of the population while working with those who have special needs.

## 0.3 EEP is an Ongoing Process

The EEP lifecycle approach is subdivided into six main areas or stages to make it easier to understand. In practice, however, many of the stages overlap as part of a continuous, reiterative process. Realistically, all you can expect to achieve is a viable solution which will always need to be kept up to date at all levels. Like painting one of the world's huge bridges, as soon as you get to the end, it's already time to start again from the beginning.

### Quick-Thinking Firefighters Evacuate Trapped Tenants

It was a cold winter night. Tenants were trapped in their apartments when a fire broke out on the main stairway of an old five-story suburban London apartment building. Dating back to the Victorian era, the building featured a common stairway shared by all ten apartments. Once that stairway was blocked by the blaze and smoke, everyone inside was prevented from fleeing to safety via the only exit.

Fortunately, quick-thinking firefighters – who were familiar with the layout of these antiquated buildings – immediately ordered the panicked residents to turn around and go back up the stairs, climb into the roof space and make their way towards the building next door, where they would meet their rescuers.

Meanwhile, two firefighters were dispatched into the neighboring building, where they climbed up into the roof space and broke through the intervening wall to allow the people to escape from their smoke filled section into the clean air, where they could clamber down and out into the street. Only a few weeks after the fire, the managers of the property had fire doors installed in the roof space to allow occupants to transfer from one building to another in an emergency. Everyone learned from this close call, and this building-to-building lateral escape method was established as a standard for similar old apartment buildings in the area.

Without the local knowledge and resourcefulness of the firefighters, the apartment fire could have quickly turned into a tragedy.

## Discussion Questions – Introduction

At the end of each section dealing with a particular phase of the EEP program, you will be asked to ponder a few points. The aim is to get you thinking about the subject as a means of reinforcing the learning. For example, you might consider what makes you interested in EEP and what you think you can contribute to the subject. This might lead to some useful realizations which may help you to focus on the particular aspects where there are gaps in your knowledge or areas where your background and skills may prove beneficial.

1. Should EEP be seen and thought of as a part of, or an extension to, normal fire safety procedures or does it include fire escape and drill procedures as an integral part of a larger whole? How are you going to make your views known, accepted, and adopted?
2. Based on your personal beliefs and experience, how do you think EEP might be affected by various aspects of culture? Bear in mind the cultural influences which may arise from directions such as national and religious identities; industrial norms and standards; trade unions and pressure groups; age and health groups; or, the culture promoted within the organization either by intent or default. Can you identify relevant cultural issues in your environment?
3. Think of three or four major emergencies that have been covered by the media and comment on the evacuations which took place. Just based on your own common sense, what would you say worked and didn't work? Are there any lessons to be learned from those events? How do these observations help you to identify what skills you would like to learn from this book?
4. Presumably, you will have taken part in a regular evacuation drill – but have you any direct experience of a real-life emergency evacuation? What, if anything, does that tell you? Do you think this will affect the way in which you approach or deliver the subject?

# Phase 1

## Set Up the Emergency Evacuation Planning Program

Over the years, business continuity (BC) planners have dealt with emergency and contingency plans for responding to threats as diverse as fires, floods, hurricanes, typhoons, tornadoes, earthquakes, terrorist activity, riots, demonstrations, and military coups. In each of these emergencies, the basic question is: What actions are best to ensure the safety of the people?

*This section will help you to:*

- ➢ Use the lifecycle model to create the methodology for EEP.
- ➢ Research and write a proposal and business case for your EEP program.
- ➢ Identify the circumstances that can be expected to trigger an evacuation.
- ➢ Be aware of your legal and regulatory requirements.

# 1.1 A Formal Methodology

## 1.1.1 The Lifecycle Model

In the world of business continuity management (BCM), professionals have used this type of lifecycle model as the basis of a formal methodology, or management system, embracing specialized tools, techniques, and terminology. This general acceptance adds a certain amount of gravitas to the whole procedure and offers a standardized way of achieving clearly defined objectives. In this book, I have developed a parallel set of tools, techniques, and terms which can be applied to EEP to demonstrate that we are pursuing a due process as we carry out our responsibility to protect those who are under our care, visiting, or using our premises.

Within this formal method, you will need to identify or determine exactly where the various associated responsibilities should, or do, lie. We will come back to this aspect in a later section of the book in which we discuss management and control.

**Your sponsor should be someone who "knows the ropes" in the organization.**

## 1.1.2 Prerequisites

- **Sign-on and sponsorship:** Before you embark on an EEP you will need authorization. The scope will need to be determined and agreed upon by top management. In addition, your program will require an identified sponsor who can provide the authority to back the work, its requirements, and implications. Your sponsor should be someone who "knows the ropes" in the organization, is skilled at communicating at all levels in the organization, and has the confidence of upper management. This is the person who is to take overall responsibility for emergency evacuation.
- **Policy:** To formalize this initial phase, you will need to have a documented policy statement which has been signed off by the sponsor. Such a policy should state clearly why the organization is embarking on an emergency planning program, referencing the drivers behind that thinking. Drivers may include such concerns or considerations as corporate responsibility, staff welfare, health and safety legislation, industry guidelines provided by a trade body, or regulations imposed by a regulator.
- **Scope:** The scope of the accompanying exercise or testing program should also be clearly defined and stated within the policy document. Elements, or dimensions, of the scope may include geographic limitations together with an indication of the type and scale of the scenarios, which the plans are expected to cover.

Ideally, the scope should also include some reference to the type and level of aftercare which should be provided, perhaps in accordance with prescribed individual needs.

- **Strategy and tactics:** With the policy in place, you will have the authority to proceed with your investigations and move on to develop the strategy and tactics which will meet the aims of the policy. Throughout the rest of the program, you should not only follow the policy but also make sure to refer to it as the source from which the strategy and tactics are derived. Keeping the policy as the reference point in your communications is a clear demonstration of good practice and provides the transparency which is expected of good governance. In the wake of a serious incident, it is almost certain that an inquiry will be held and evidence of due care and attention could become a serious issue. In a court of law or a formal inquiry, the lack of such evidence could be interpreted as negligence at the very least. That negligence would be laid firmly at the feet of whomever is deemed to be responsible for emergency procedures, and that will include you and whomever you report to in this connection.

## 1.2 Program Management

Because EEP is a long-term program rather than a short-term project, it needs to be approached and managed in a manner consistent with the long life that is to be expected. You and other people will need to produce a series of deliverables, or documents, at milestones along the way. One of the inherent difficulties in this type of work is the tendency to wander off course due to lack of attention to detail, misunderstandings, or forgotten intentions.

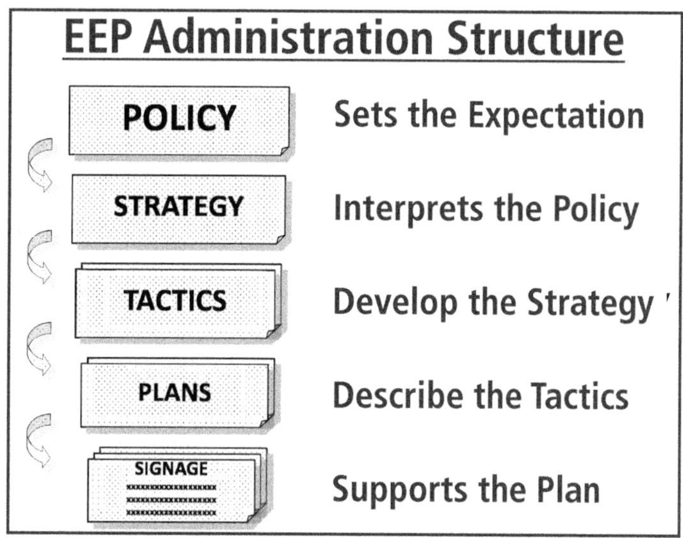


Note that in Phase 1 of this methodology, you will not be completing the entire planning and implementation process. Rather, you will be setting up the administrative structure and the milestones for creating and implementing the EEP

– and you will be prepared to explain the proposed process to management as you seek support for this initiative.

**Administration structure:** To stay on course, we advise that you set up a workable administration structure right at the outset, a structure complete with a set of properly documented reference material outlining who does what, why, when, and how.

A typical administration structure has five levels of detail, each with its own specific purpose and characteristics. Working from the top down, you would normally start out with a *policy*, from which you develop *strategy, tactics, plans*, and *instructions* for carrying out the plans (in the case of evacuation plans, *signag*e would be one example of conveying instructions).

As a general rule, you will find:

- Directors (or their equivalent) are the people in the organization who work with *policy*.
- Directors allow, or authorize, managers to work on their behalf with the *strategy*.
- Managers, in their turn will expect the supervisors who report to them to work on the *tactics*.
- Supervisors will direct the work of individuals, who will work with the development, delivery, and maintenance of the actual *plans*.

Now let's take a closer look at each of these documents and how it is developed. You could agree to all of the details in theory without necessarily committing your thoughts to paper. However, that could lead you into all sorts of difficulties. Without good administration, it is often difficult to recognize success because we simply don't know how to measure or evaluate it. In fact, without a clear definition of what we are trying to achieve, our success is not even recognizable and our hard work won't get acknowledged, which can be rather discouraging to say the least.

**Management at the highest level of authority should sign off on the policy, expressing the clear intentions of the organization in regard to the health and safety of human beings in the event of an emergency.**

## 1.2.1 Policy

To begin, a formal policy is the only way to ensure that you are working to a corporate statement of expectations. Management at the highest level of authority should sign off on the policy, expressing the clear intentions of the organization in regard to the health and safety of human beings in the event of an emergency.

The policy statement points you in the right direction, tells you what your goals are, and provides you with the authority to proceed and engage the attention of others. In addition, when you meet resistance from your colleagues, a formal written policy allows you to make clear that your requests are not made in the context of some personal project but in the service of a high-level corporate initiative. Ideally, the policy should be brief and relatively non-specific so that it can be easily interpreted to fit a variety of situations. Oddly enough, a document which is too elegant and precise may actually work against you because its very precision will prevent you from adapting to all the countless variables which you might come across.

## 1.2.2 Strategy

At this point, it is important to review the difference between a policy and a strategy. While your high-level policy is a statement of intent that sets out the commitment of the organization, that policy provides no detailed guidance on how those intentions are to be met. It is the strategist who turns those high-level intentions and ambitions into a workable strategy that's both affordable and effective. Thus, strategy is a practical and viable interpretation of the policy which provides a broad outline of how those ambitions might be achieved. For example,

- Your *policy* might say: *We regard the safety of all personnel as paramount.*
- Your *strategy* would be rather more specific, saying something like: *We require emergency evacuation plans (EEPs) and procedures to be developed, tested, and maintained, sufficient to ensure the safety of all personnel.*

Your strategy needs to explore the various kinds of emergency situations that might occur and how best to react to ensure the health and safety of people. It is important to discuss and compare the various possibilities for ensuring human health and safety – taking into account their feasibility, effectiveness, and potential costs – before a final EEP approach is decided upon and endorsed by senior management.

Inevitably, questions will arise as to costs and timescales of implementation; effectiveness of the resulting procedures; degree of associated risks; and compliance with relevant standards or regulations. You should be prepared to offer opinions or provide answers as the work on the development of the EEP strategy proceeds. This degree of preparation means scanning the horizon and keeping abreast of developments within the working environment as well as understanding the needs and expectations of your colleagues.

## Case Study: Evacuation by Lucky Escape

Over the past decade, Excelsium Insurance,* a 200-employee insurance company, had implemented a typical business continuity program for a company of that size. The company was located a few hundred miles north of the region in the central United States known as "tornado alley," but it had never suffered from storm-related threats and vulnerabilities. While management was aware of a potential tornado threat, the danger had never seemed "real" – so near and yet so far – and the business continuity program had not addressed it in any kind of realistic detail.

The company had what might be called a one-size-fits-all evacuation plan, conducting all-purpose fire drills for their seven-story headquarters building two or three times a year, evacuating all workers to the parking lot. In practice, the drills were not taken seriously. Floors 1 through Floor 6 always evacuated; employees looked forward to the break. Significantly, Floor 7, the executive level, never did participate, with executives chuckling, "We keep getting these warnings, but even the old-timers can't remember a tornado actually hitting this town."

Over time, employees viewed these exercises as nothing more than a convenient excuse to play ball – in fact, they kept a sack of soccer balls handy and looked forward to the "exercise." When Matilda, an otherwise soft-spoken, mid-level accounting manager who hailed from Oklahoma and had lived through many tornados, wrote a strong memo to the executives expressing her concern that the parking lot was probably the worst place to take refuge in a real tornado, they laughed and called her a "sissy."

One early spring day, the weather service issued warnings that a large tornado had been spotted in the area. The alarm sounded; tornado evacuation procedures were initiated. A darkening sky suggested that this was not a false alarm. Within minutes, confusion reigned. Suddenly, it was no longer soccer-playing time. Panicked employees quickly recognized that the parking lot was not the place to be as the tornado approached – but it was also clear to them that there was no other place to go. Employees plunged down the stairs toward the exit, only to run headlong into other employees coming back in. Executives looked out the windows and saw the signs of impending doom. They wisely chose to leave the seventh floor and participate – but even they had no idea what to do.

Nobody was in charge. The well-defined and now obviously incompetent all-purpose evacuation procedures were immediately forgotten. In the midst of confusion, seemingly out of nowhere, Matilda realized nobody else was stepping up to the plate. With a storm approaching and lives at stake, she was soft-spoken no more, drawing on a lifetime of tornado drills in school and at home from growing up in Oklahoma. Very aware of what steps to take in the face of a tornado, Matilda immediately issued commands in a clear, confident voice, quieting and calming her co-workers – some of whom were at risk of being trampled – and quickly and effectively directed employees to relative safety in the lower-level stairwells.

As luck would have it, the tornado veered away and did not come within a mile of the building. If it had come much closer, the impact would likely have been catastrophic. While there were some injuries, none was serious. After that close call, Matilda was invited to the seventh floor to review her earlier memo with all the executives. Today, Excelsium Insurance no longer trusts to luck and an all-purpose evacuation process. The company now has a tornado shelter and fully exercised facility evacuation plan.

* *This case study and others are based on actual situations in real enterprises. The names have been changed and, where necessary, some of the details may have been modified in order to maintain the confidentiality of the enterprise.*

## 1.2.3 Tactics

Tactics are derived from the agreed strategy; it is at this level that you will be developing realistic project plans regarding the ways and means to effect an emergency evacuation.

At the tactical level, you will need to apply your technical knowledge and provide guidance to those above you in the organization. At the same time, you will be expected to work closely with your peers who will probably have a rather down-to-earth pragmatic view of what might go wrong and how people might react. Much of their knowledge and opinions will be derived from practical experience which is, after all, the best source of information regarding what is possible and how best to deal with problems and eventualities.

In contrast to the high-level policy, which is applied to the organization as a whole, the strategy may apply to particular locations or divisions, and the specific tactics would apply to specific sections or areas where people might be found. Thus, plans themselves are aimed at the level of individuals or groups of individuals, while signage is an example of a tactic that is there to support individuals as they find their way along the route towards safety.

A detailed risk assessment and an exploration of the environment carried out together with those who are familiar with the business operations and the facilities will provide you with much of the input you require at this stage. It is normal to work out the tactics for the most critical area of the business operation as the first priority. Later on, the tactics for other areas can and will be developed in the light of the experience gained. While the first set of tactics may provide a basic model which can be adapted or adopted for other areas, it is important to retain a degree of flexibility in your thinking in order to be able to recognize and employ the tactics which are most appropriate to the situation, the building, and its occupants. In other words, don't get "stuck in a rut" and try to avoid "tunnel vision."

**A detailed risk assessment...carried out together with those who are familiar with the business operations and the facilities will provide you with much of the input you require at this stage.**

While thinking about tactics, you will be developing a strong feeling for the type and level of plans which might be required. Factors to be taken into account include the complexity of the environment; the likely escape routes; the variability of occupants; and the range of their capabilities in regard to mobility and comprehension of messages and signals.

An important question at this stage is how you will develop and distribute the actual EEPs.

- **Will there be a generic plan which is relatively simple to communicate?** In that case, it may need only to be seen and understood by those who would be leading and managing the evacuation. Such a plan would require a limited distribution and publication process.
- **Will there be a number of similar but distinctive plans for different locations or areas?** These differing plans would imply that more people will need to be given access to, and develop a full understanding of, the EEPs and procedures which apply to their areas or departments.
- **Will there be a variable or transient population?** If so, there may be a need for individualized or personal emergency evacuation plans (PEEPs). This situation implies the need for the development and distribution of templates and providing support for those who need to customize or adapt plans to suit their specific needs.

Tactics include establishing the way in which the various types of plans will be developed, delivered, customized, and maintained.

## 1.2.4 Plans

It is at the planning level that you and your colleagues will need to develop the specific plans, resources, facilities, and techniques to get people from their expected location to a place of safety in the event of an emergency.

You will need to make arrangements to devise detailed plans of action for all types of people to use as the basis of their evacuation in the event of an emergency.

If the emergency evacuation procedures are relatively simple and common to all parts of the premises, then you may require only one generalized EEP which covers all eventualities. Use of a single plan implies that the premises are probably designed with safety of the occupants as a key requirement and that exit routes are straightforward and obvious. The prudent use of suitable signage will also be helpful in such circumstances.

**At this stage you will have, or should have, developed a thorough understanding of the environment and the people who are likely to be present at any one time.**

However, in the majority of cases, the premises will offer sufficient variety in the possible routes to safety meaning that rather more detailed, varied, and specific plans will be needed. To meet this need, you will be required to think about the use of suitable templates, which can then be used as the basis for developing a suite of specific plans, each of them dealing with a particular situation, circumstance, or type of person.

Clearly at this stage you will have, or should have, developed a thorough understanding of the environment and the people who are likely to be present at any one time.

- At the *tactical* level, you will have decided on the range and type of plans which need to be generated.
- At the *planning* level, you will need to organize and arrange the development, writing, publishing, and delivery of such materials, which is where the bulk of the work will need to be done.

However, much of this work can probably be delegated to those responsible for a particular group, section, or person. Your principal role will be that of leader and center of competence with the responsibility for ensuring that a complete range of suitable plans emerges from the process.

Tactics will determine that these plans should be exercised and tested at some stage and this type of activity will lead to their refinement and improvement in due course. Exercising, testing, and the associated procedures will also spread the word, boost confidence, and enhance competence among the target population.

## 1.2.5 Signage

The requirements and expectations with regard to signage will vary according to a number of factors, such as the nature of the organization; the size, type and structure of the building; its potential occupants; and the location of suitable safe assembly areas.

The *purpose of signage* is to provide guidance and explanation; this is made easier by adopting the common language of standard signs as described in "Signs and Signage" in Phase 4, later in this book. In an emergency situation, the messages need to be simple and clear so that anyone can understand them even in stressful circumstances.

In most buildings, wherever there is some degree of public access, there will already be signs in place to indicate the fire escape routes and fire exits. Such signs are usually a result of a regulatory or a statutory requirement. However, you should make a point of checking that the signage is actually there and that it provides an adequate level of information to all those who may be present in an emergency situation. Key points to look out for include:

- A clear indication of safe assembly areas.
- Consistency of indications along the exit routes.
- Exit points clearly identified.
- Emergency access features clearly explained, e.g., how to open fire doors.
- Adequate lighting where it might be required.
- Sufficient capacity in passageways and assembly areas to deal with the largest anticipated volumes of people.

## 1.2.6 Program Management Viewed as a System

If the EEP program is to be a long-term re-iterative process then, perhaps, you will find it useful to regard it as a type of management system. In the world of standards, the various national and international bodies, such as the British Standards Institution (BSI), the International Organization for Standardization (ISO), and the American National Standards Institute (ANSI), all use a specific *lifecycle* model as the basis of any management system.

This model is known as the Plan-Do-Check-Act (PDCA) approach, which we should adopt, or refer to, as an over-arching model for our ongoing EEP management system. PDCA is shown here in graphic form. As you can see, it is a continuous reiterative process with four main stages or phases. Intuitively, you would expect to start at the top and work around the cycle with each stage leading naturally on towards the next one which deals with the outcome from its predecessor. Eventually, you return to the original starting point and repeat the whole process again and again.

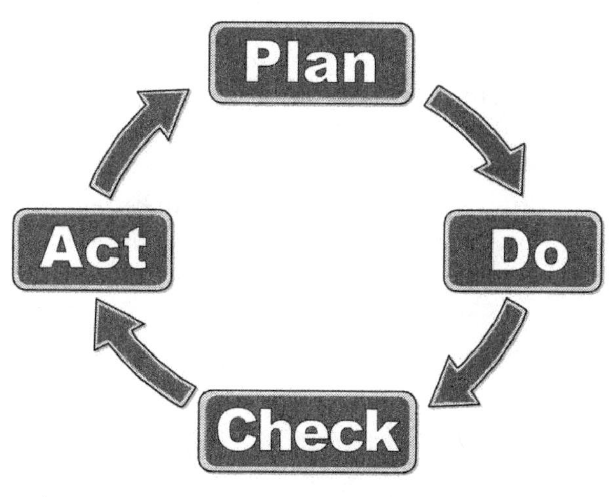


Let us take a deeper look at how the four steps in this PDCA approach work in practice.

1. At the outset we *plan*. In this context, that means thinking ahead and planning what needs to be done and how it will, or should, be done. Such planning might include reference to any standards which should be adhered to or regulations which should be taken into account. The result, or outcome, of this initial stage will be a clear indication of the expected program of work and what it is expected to achieve or produce.
2. Once we know what is expected of us, we can start to *do*. That means that we carry out the tasks or actions which were listed in our initial plan, following any suggested guidelines, and working within the standards or regulations which are deemed relevant. Such actions will lead to the development, production, and delivery of a product or service which represents the true purpose of this management system.
3. The third stage is to carry out a *check* to ensure that the outcome meets our expectations. This check is mostly a measure or test of the suitability of the process and it should reveal any shortcomings within the system. The assumption here is that a perfect system should produce a perfect product and that an imperfect product is the result of an imperfect system. The outcome from this stage is either a verification that the system is fit for purpose or a set of actions which will address some or all of the shortcomings.

4. Stage four is the *act* stage which is where we carry out the actions that are required for improving or enhancing the management system. The outcome of this stage will be a better management system which we can then use to continue to deliver an improved product or service.

Over time, this approach will allow us to monitor and improve the underlying management system because it is a continuous self-correcting system. It will also adjust itself automatically to take into account any changes in the operational or regulatory environment which may occur over time.

You should bear this PDCA model in mind when setting up and running your EEP program.

## 1.3 Policy Development and Management Approval

As we have already said, you do need to gain the full support of top management from the very start of the program you are about to embark upon. This means settling your terms of reference (TOR) and underpinning them with a clear policy.

**The policy document will provide the framework around which the EEPs and capabilities are designed and built.**

The policy document will provide the framework around which the EEPs and capabilities are designed and built. Its value stems from the fact that it can be seen to be set out and owned by top management. Its purpose is to communicate to stakeholders the principles to which the organization aspires and against which its performance can be measured or audited.

Because the primary purpose of the EEP policy document is one of communication, it should be short, clear, precise, and to the point. A lengthy and complex policy will undoubtedly prove to be a barrier to communication.

As a minimum, this policy will identify the following elements of the program:

- Objectives.
- Scope.
- Responsibilities.
- Methods and standards.

The policy development process starts off with you identifying the key components, which will include a definition of what is meant by EEP. This definition has to be one which reflects, and is relevant to, the organization concerned. Therefore, it has to be developed and agreed upon in-house, rather than simply lifted off the shelf or copied from a standard or reference document.

Other key components will include any relevant standards, regulations, and legislation that must or should be referenced together with good practice guidelines or policies from other sources which might be viewed or used as a benchmark. Perhaps the most important component will be a definition of the purpose of the EEP program, i.e., what you are hoping to achieve.

Once you have prepared a draft policy, you should circulate it among the top management for consultation. Then you will need to finalize or amend the draft document in accordance with the feedback you receive before getting it signed off by top management. At the same time, you should obtain management approval of your intended strategy or TOR for implementing the EEP program which stems from the policy.

Finally, you will be ready to publish and distribute the EEP policy using the version control system and techniques which are appropriate for your organization.

## 1.3.1 A Practical Approach to Developing Policy

Developing an EEP policy need not be a particularly complex or lengthy process, but it might be useful to carry out a little research before you start to get some idea of what is needed and what is possible. For example, there are probably a number of policy documents in existence within the organization which give an indication of the style, layout, and content that your top management might expect to see. You might also be able to refer to external resources for guidance, but at the end of the day, common sense and knowledge of the culture and style of management should enable you to develop a reasonable draft policy.

Basically, your draft policy needs to state what you are trying to achieve and the reasoning behind that objective, such as staff welfare concerns, and this argument needs to be linked to the organization's aims and ambitions as expressed in the mission statement.

The policy should include, or refer to, the organization's definition of EEP and how it is to be organized and managed, together with clearly defined responsibilities. A clear definition that is embedded within an approved policy document will give you the authority to proceed. Without such clarity, the scope and purpose of the whole program can easily become a subject for debate whenever you are seeking assistance or cooperation from other busy people.

Like all policies, this one should be reviewed on a regular ongoing basis and whenever there is a significant change in the working environment which may affect the evacuation procedures and plans. Changes in legislation or regulations in regard to health and safety or personal welfare may also trigger the need for a policy review.

Once you have prepared your draft policy, you will need to get it signed off by the sponsor or project champion who should be someone whose rank and position will command respect from all those whom you will be working with. Obviously, this person should be selected and cultivated before you start to develop the policy. The input, advice, and guidance of this person will help you to get off to a good start with the creation of a policy which is fit for purpose within your particular working environment.

You will find a couple of examples of EEP policies in the accompanying EEP Toolkit which you can adopt or adapt for your own purposes. If you should decide to use one of these documents as a template, remember to ensure that the definitions, intentions, objectives, and any underlying assumptions are appropriate for the size, type, and nature of the organization you are working with or for. When discussing the draft with your potential champion ask him or her specific questions in this connection, such as: "Are we making any assumptions here which could prove to be false or misleading?"

The *Brief EEP Policy* example in the accompanying EEP Toolkit is a straightforward policy statement which simply states the organization's intentions with regard to EEP. You will also find a copy of an *Extended EEP Policy*, which includes an extension that describes the lifecycle model which the directors expect to be used as the reference model for the EEP program. This extended version may be more appropriate for use in those organizations which expect, or require, full details of the intended process or procedure to be included in, or attached to, their policy statements. The choice is yours.

### 1.3.2 Gaining Support

As you try to win support from your colleagues or superiors, you must be able to deliver your argument effectively, meaning you need to use the appropriate language, provide a good reason for supporting your cause, include the right level of information, and be prepared to handle any questions which may arise. The other requisite for success is genuine enthusiasm for the subject. So it helps if you can start by explaining why and how you first became engaged with the subject. This approach develops empathy and shows your listeners how they might come around to your way of thinking, especially if you were attracted by a rational reason such as the associated needs and benefits. Before confronting anybody with your argument, you need to marshal your thoughts and ideas together so as to make the most effective use of them.

I urge you to think about why you want to proceed with this emergency evacuation venture. Introduce the subject by explaining how and why you came to believe in it, even relating a true story that you or someone you know experienced personally. This narrative can lead naturally into a simple explanation of what is involved in keeping everyone safe in an emergency and what is to be achieved. Once you have dealt with their questions and everyone has an understanding of what you are asking, you can move on to tell them about the benefits which might include compliance with regulations or legislation, improved morale, a better public image, and support of the corporate mission or the organization's intentions with regard to the community and the population.

Sometimes it helps to mention the disadvantages of inaction, which might include the possibility of an enquiry in the event of something going very wrong in an emergency evacuation situation. Remind your listeners that people who are injured, inconvenienced, suffer loss, experience trauma, or otherwise consider themselves to have been ill-used in the evacuation will give detailed and dramatic interviews to the press, keeping the story of the company's botched evacuation going for days. Additional negative publicity will follow if someone sues for negligence – this can impact on the organization's brand and image negatively for years. Subsequent recovery of the reputation could be a long and difficult process which may never fully succeed.

**Give them a choice:** Inevitably, at some stage your potential champions will want to know about the likely costs involved in this effort. You should be prepared to give them some approximate figures in relation to expected capital expenditure, the workload involved and the anticipated time frames for the various stages of the implementation.

Wherever possible, I like to give those to whom I am making the presentation several alternative approaches from which to choose. If you offer people just one option, then their answer is a straightforward choice between "Yes" and "No." On the other hand, if you give them a choice of several solutions, "No" becomes a relatively unlikely response. My preferred approach is to offer them three choices, giving meaningful labels or nicknames to each choice to distinguish them more easily and to capture the attention of the audience. The three choices are.

- An over-the-top elegant or magnificent solution, which I might call "Elegant but Expensive" or "Leading Edge."
- A down-to-earth, practical or pragmatic solution, which I might call "Best Value."
- A rough and ready basic solution which barely meets the requirements but has the potential for enhancement at a later stage, calling it perhaps "Cheap and Cheerful."

## When Evacuation Problems Hurt Company Reputation

### *Herald of Free Enterprise Disaster*

On the night of 6 March 1987 a roll on/roll off car ferry on the Dover-Calais route, MS Herald of Free Enterprise, capsized off the coast of the Belgian port of Zeebrugge bound for Dover on the south coast of England. Of the 80 crew and 459 passengers on board, only 346 survived. Most of the casualties were people who were trapped inside the ship and succumbed to hypothermia due to the cold water (3 degrees C/37.4 degrees F).

The ship was operated by Townsend Thoresen on behalf of Peninsular and Oriental Steam Navigation Company (P&O), which had purchased the company in 1986. As part of the public inquiry following the disaster, Lord Justice Sheen wrote in a July 1987 report that Townsend Thoresen possessed a "disease of sloppiness" which permeated the company's hierarchy.

The cases surrounding the incident set a precedent in English law that corporations could be prosecuted for "corporate manslaughter." The jury at the coroner's inquest found the company guilty of manslaughter, and seven employees were charged individually with manslaughter. While none of these individuals was actually convicted, the ship, which was able to be re-floated, finally had to be sold for scrap since nobody wanted to buy her.

P&O took action quickly to repair the damage to reputation from months of negative world-wide publicity. In October 1987, P&O rebranded all the operations of Townsend Thoresen with a new name: P&O European Ferries. In the rebranding process, P&O had all the red hulls in the fleet repainted navy blue and the familiar TT logos removed from the funnels. Thus, P&O managed to save its ferry business by obliterating all vestiges of the recognizable Townsend Thoresen brand. Eventually, P&O dominated that sector of the market again.

To begin, I argue quite strongly for the elegant or magnificent solution. I then climb down to the sensible or pragmatic one which I describe briefly before moving on to the basic rough and ready proposition which I describe in detail together with the list of potential enhancements which can be implemented at a later stage. Usually, they want to go back to look at the "pragmatic" solution, which seems to them to be the best compromise, especially if the list of improvements is extensive and arduous. Of course, it is quite possible that they may actually choose the "magnificent" solution – which I regard as a bonus – although "pragmatic" middle-of-the-road choice was, and is, my objective.

## 1.3.3 Terms of Reference

If your company is like most sizeable and well-regulated organizations, any new project or program will be subject to a set of rules and conditions, a kind of project road map, normally known as the *terms of reference* (TOR), sometimes called the scope of work. Such documents take many forms; they can range from a simple statement describing the responsibilities and expected outcomes or deliverables up to an impressive form of legally binding contract specifying all aspects of the project or program in minute detail together with appropriate penalty clauses. Obviously, you will be aiming for something in between.

The advantage of having the TOR agreed to and documented in advance is that you will have the go-ahead to proceed with the work without constantly having to ask permission to take the next action or needing to beg for funds. In other words, the TOR can be seen and used as the authority to proceed, providing you stay within the bounds described therein. Clarity is therefore required in order to be certain that your activities and intentions are perceived as above suspicion and beyond challenge.

You will always encounter at least a few people who do not subscribe to the need for the level of detail, or even the EEP program itself, and will go out of their way to avoid involvement with this program. It is always best to be prepared to counter their arguments, or engage their support, through proper authorization. If you do find you are able to proceed without any counter-intentions or awkwardness to slow you down or divert your attention, then it is probably because you are able to adopt a sincerely confident approach based on the knowledge that you are fully prepared for any eventuality.

As we have already suggested, TOR can take many forms, but once again, we would always advise that you tend towards practical simplicity and the bare necessities rather than sophisticated elegance and complexity in what is, after all, a basic document. Here we are using the term basic in all of its senses: simple, fundamental, and easy to understand. You should be looking at a document which will provide the foundation or basis upon which you will be able to build a successful and effective EEP program together with its products or outcomes. In other words, this document should be in keeping with the rest of the EEP program – user friendly and fit for purpose without any unnecessary frills or embroidery.

**Minimum TOR:** At the minimal end of the scale, the TOR should at least state:

- The objectives or purpose of the program.
- The scope.

- Who is to be involved.
- The expected outcome(s).
- The activities required to achieve those outcomes.
- The anticipated completion or delivery times.

**More detailed TOR:** A rather more comprehensive version may also include such items as:

- The background and the aims of the program or project.
- A description of the expected deliverables or products.
- The assumptions upon which the work is to be based, together with any constraints which are to be imposed.
- Reference to any standards or regulations which need to be observed or taken into account.
- Guidance on the management controls or reporting procedures which are to be adopted.
- The audit and/or review process which is to be adopted.
- The procedure for resolution of any issues which might arise. Issues to be resolved might include such matters as technical, philosophical, practical, and budgetary considerations.

In either case, the TOR should conform to the organization's normal style in regard to formal documents which are likely to be used for reference, management, control, or audit purposes. They should also comply with the organization's document control policy.

Together, we are probably looking at about a dozen statements or paragraphs which will provide you and your colleagues with details of the tasks and how, when, and why they are to be tackled.

In the EEP Toolkit, you will find a couple of sample templates for TOR which are based upon the types of documents used by a large international membership organization. The first one is used for what the organization regards as minor or short-term projects and the second is used for major or long-term projects. They are entitled Minor TOR and Major TOR, respectively. You can choose the one which is most suited to your needs or, perhaps, you may be able to find and use a template which already exists within your working environment.

Whichever starting point you choose, you will be wise to populate the document as a draft and discuss it with your sponsor or champion before creating a final version to put forward for endorsement, i.e., approval and sign-off by top management. Normally, I would expect to put both the policy and the TOR forward for endorsement at the same time as they are inextricably linked as part of the same authority and game plan.

Sometimes an organization will also expect to see an action-tracking document as an appendix to the TOR. Such a tracking document usually takes the form of a spreadsheet which lays out the milestones and responsibilities for the program or project with space for comments and indications of progress and completion of the various actions. If your organization subscribes to this approach, then I am sure you will already be familiar with the action-tracking system which you will be using.

## 1.4 Management Approval

No matter how important, urgent, or necessary any program of work might be, you will not get very far without the approval and support of senior management. Here we are using the generic term senior management to describe those who control budgets and take responsibility for running the organization. In practice, they may be known as owners, directors, executives, stakeholders, or whatever title is appropriate to the organization and culture. No doubt you will be able to interpret the term and can identify who those people are within your own hierarchy.

### 1.4.1 The Initial Approach

You have two basic possibilities when it comes to setting up an EEP project:

- The first and ideal situation is one in which *top management realize the need* for themselves and select someone like you as their appointed representative. In that case, you simply have to follow the natural path towards the destination which they have set for you. Of course, it is quite possible that the final destination and the route may change as some of the previously unknown facts and options begin to emerge and all of you gain more clarity on the subject and its implications.

- The second and more likely situation is one in which someone like yourself recognizes the need for something to be done and *you have to win the support and approval of top management.* Under these circumstances, you will need to learn the rules and develop a game plan before setting out on your adventure trail. Think a few steps ahead: what are their possible or likely responses, and how will you deal with each of them? Don't let yourself fall into the trap of not knowing what the next step should be, just because you hadn't anticipated that particular response, result, or outcome.

**You must also think about who is going to be able and willing to open doors for you or point you in the right direction.**

In either of the above scenarios, before you go in front of top management, cap in hand, you need to develop a clear idea of why this project has to be started; what it will achieve; how and when it will produce the desirable outcomes; and what the cost will be in terms of time, labor, resource, and finances. One of their requirements will almost certainly be a cost/benefit analysis – they will want to know what they are getting for their money. Are there any hidden or subsidiary benefits which you can show them?

You must also think about who is going to be able and willing to open doors for you or point you in the right direction. It isn't all about *what* you know; *whom* you know (or have access to) is of equal importance.

In a large and complex organization, it might be prudent to consider a number of people as possible supporters, each perhaps with a different angle or position within the hierarchy. You could end up with someone appointed at the very highest level to act as the policy maker or sponsor. Then there may be someone who is made responsible at the strategic level who would be acting as your champion. At the tactical level, you may also require the support and assistance of a number of managers, such as the heads of personnel or HR, health and safety, facilities, and security. If there is an internal training department, they may also be useful as a source of ideas and help.

## 1.4.2 Bidding for Permission

Once you have a clear idea of who might prove to be useful supporters in your campaign, you need to prepare to make a bid for permission to initiate and implement the establishment of an EEP project. While we realize that this is to be the start of an ongoing program or reiterative process, it might not be necessary to draw their attention to the long-term requirements at this stage. Focus on getting permission to get something up and running, leaving the longer term considerations and aspects until later. Over time, their support and commitment will grow. Once they see and appreciate the products and benefits of the program, they will gradually come to accept the need for some ongoing maintenance and updating.

You need to be prepared for two aspects of the bid. First, you have to convince them that something should be done. This is about why they must engage in this program and what they will get. It is partly a question of costs and benefits, which is the logical business case, and partly penalties and advantages, which is the philosophical self-protection aspect.

The second half of your argument has to be about the practicalities, which means letting them know how it is to be achieved. You will need to have some sort of draft documents ready for their review and subsequent approval. Here I am thinking about a draft policy statement and an outline of the proposed strategy (subject to review and confirmation after the initial authorized investigation). The point is that you will need to strike while the iron is hot. If you manage to convince them that this is a worthwhile project, you need to get their authorization to proceed before they change their minds or forget. I still recall the surprise I got when my first potential EEP customers changed their minds about the need for proper evacuation plans only a couple of months after their involvement with one of the world's worst tragedies, the World Trade Center in 2001.

Obviously, the details of how you arrange and manage this stage must reflect the nature, scale, and infrastructure of the environment in which you work, but if, as I assume, you work in a typical organization, perhaps you can adapt my thoughts to fit your circumstances.

### 1.4.3 The Basic Argument

Once you have made it known that you are serious about the subject and have a reasonable idea of who might offer you the backing you need, you will need to make an appointment to spend some time with potential supporters sharing your views and arguing the case for setting up an improved EEP program. I would assume that some form of evacuation procedure is already in place under the guise of the fire escape procedures which are a legal requirement almost everywhere. What you are hoping to achieve is something with a little more substance which will cater to a wide range of possible emergency situations, many of which will not be fire-related.

I would expect that you need to bid for an opportunity to be on the agenda to raise the matter at a regular meeting at which your top management will discuss issues regarding policy and its strategic implications. In the typical business set-up this might be a board meeting or an executive council meeting. The likelihood is that you will be assigned a brief slot towards the end of the meeting, which means they have already discussed a few important issues and will be looking forward to closing the meeting. In other words, you must be brief and get to the point straight away.

Once you have explained in broadest terms what you want to achieve and why, you need to be prepared to offer them some alternatives so that they can make a choice. My normal strategy is to offer them at least three choices, which reduces the chances of them saying "No" (a likely result if I offer them only one option).

**One volunteer is worth a dozen conscripts.**

If I am asked to suggest who should be nominated as my sponsor or champion, I will choose the one who is closest to me in terms of rank and position. I will also be quite prepared to accept the offer of someone from a higher position who might come forward as a volunteer. One volunteer is worth a dozen conscripts. When the "top dog" comes forward in this way, it is likely that my original suggestion will be nominated as the one I should be reporting to on a regular basis. In this case, I will have gained not only a sponsor who would be responsible for authorizing the policy in this regard, but also a champion who will ensure that the strategy is carried through to meet the expectations of the policy.

## 1.5 Making the Business Case

Before a business organization is likely to pledge support in terms of funds and resources, the key decision-makers will need to be convinced that they are going to get a worthwhile return for their investment. They need to be convinced that they will be getting value for their money and that the outcome is in line with their corporate ambitions. The task before you is to make and present a sound business case.

This process is about winning them over so they will feel that they can back you with the necessary funds and resources.

### 1.5.1 Create a Strong Business Document

While there is no standard formula for making a business case, some common characteristics are shared by most key business documents:

- The overall style should be short, sharp and to the point, written in plain language, and free of jargon or complex terminology.
- It should start with a short summary or overview which the busy director can use as an indicator to judge whether this subject is worthy of his or her attention and to determine whose area of responsibility it might be.

- It is usually easier to write the summary after the rest of the contents have been pulled together as a form of *précis*. However, some people prefer to prepare the summary first – as a kind of guideline – and then go on to expand on their thoughts adding more detail as they go.
- Typically, a successful business case states clearly the problem which is to be addressed, or the opportunity which is to be exploited, followed by an outline of the proposed solution to the problem or a stratagem which should be adopted to take advantage of the opportunity.

**Outcomes...would include...a safer working environment, improved welfare, enhanced morale evidence of good governance, proof of corporate citizenship, and compliance with regulations.**

- Once the intention, and the reasoning behind it, has been made clear the business case should focus on the outcomes which in our case would include such benefits as: a safer working environment, improved welfare, enhanced morale, evidence of good governance, proof of corporate citizenship, and compliance with regulations.
- However, these are relatively intangible considerations; in addition, the business case will need to describe specific tangible products.These might include, for example, identified and qualified exit paths, escape routes and safe assembly areas; published, tried, and tested generic EEPs and procedures; published, tried, and tested PEEPs for all those with special needs.
- Your business case should present a brief overview of how the desired outcomes are to be achieved and the necessary commitment that will be required, expressed in terms of the resources needed and the key milestones which will be represented by specific deliverables. These deliverables would include such items as a risk report, proposals for access and egress improvements, agreed emergency assembly areas, plan templates, draft plans, evacuation training, evacuation drills, and tested plans.

## 1.5.2 Establish the Costs and Benefits

Once they have understood the problem, the solution, and the requirements, managers will want to know about the costs and the resulting benefits. Balancing costs against benefits is usually a question of the capital outlay and overheads involved compared and contrasted with the expected improvement of market share or future profits. In the case of EEP, the benefits are rather more nebulous but are literally of vital importance to all those who inhabit, use, or visit your premises which includes the decision-makers themselves. This last point provides a specific and personal answer to the "what's in it for me?" question in regard to the business case which we are putting before them. An individual's own personal safety is always an influential, possibly subconscious, factor to be taken into account when facing any important decision; thus, you should make sure that it is mentioned, but don't make too much of an issue out of it. Exaggeration or overemphasis could be misinterpreted as an attempt to pull the wool over their eyes, frighten them into agreement, or even the result of some obscure hidden motive. Any of these would be a distraction and a possible obstruction.

*At the end of the day the benefits of EEP should easily outweigh the costs.* This is largely because the costs should be relatively minimal and the benefits are well worth having for a comparatively small outlay. The only occasions in which the outlay is likely to be significant are those with particularly high risks, a condition which only hindsights the need for something to be done without delay or prevarication because lives are at risk.

It is quite likely that your organization will have a template or a set of guidelines for use when preparing a business case, and this will give you a good idea of the layout, style, and contents which you will be expected to adopt.

The basic elements which you need to get across are:

- Requirements or objectives, i.e., what needs to be done.
- Drivers or reasons why, i.e., why it should be done.
- Outcomes or products, i.e., how it should be done.
- Timescales or delivery dates, i.e., when it should be done.
- Implications or costs, i.e., how much it will cost.
- Penalties or fines, i.e., what inaction might cost.

Other ideas which might be worthy of your consideration when preparing the business case could include the answers to such questions as:

- Why should we spend time and money on this? This might include referring to such issues as regulations, legal obligations, and the nature of the population involved.

- Public perception, staff loyalty, occupant's safety. This might include reference to evidence of any concerns which have been expressed or events that have occurred.
- Link to existing policies and strategies. This might include reference to things like the corporate mission, personnel policies, and sales or marketing messages.
- External viewpoints of the organization or its venues. This might include the likely feelings or reactions of members of staff, customers, and the community at large. A well-managed project might expand the range and type of visitors to the venue.
- Special needs of groups or individuals. This might cover all those who do or may occupy or visit the premises.
- What our competitors do in this regard. This might be particularly relevant where health and safety or the welfare of the occupants or visitors is a high-profile aspect of the organization's aims and duties.

You might also like to reflect upon the answer to the question, "What triggered your thinking about the subject?" In other words, "Why are you bothering to raise this as an important issue?" If you can communicate your motivation with conviction, you may well win the day.

## 1.6 Managing EEP

There are two aspects of management in relation to EEP:

- First, there is the *ongoing management of the overall program* which includes development of the procedures, installation or adaptation of resources, distribution and awareness of plans, training and exercising.
- Second, there is the *occasional need for someone to be prepared to manage the response to an incident* which includes the need to evacuate the premises.

We can think of these two areas of responsibility as routine plan management and impromptu event management. Passive and active is another characteristic which distinguishes them. Apart from the obvious difference in the timeframe involved, there is also a clear distinction in the way these two roles need to be played. Role requirements may well have a bearing on the type of person who should be asked to take on the task of management within EEP.

The long-term management of the overall program is largely a series of administrative tasks, most of which are fairly repetitive. In this capacity you will need to be patient, persistent, and careful. You will need to maintain, and work to, a proper schedule in order to ensure that the activities within each phase of the lifecycle are all carried out and checked in the right sequence and at the right time. It is also important to maintain transparency throughout the whole program, which means keeping clear and accurate records so that auditors can trace exactly what happened and align it with the policy and standards to which the organization is supposed to be adhering.

Ideally, over time, you should become the center of competence for EEP which implies being capable of delivering or facilitating the various tests, exercises, and workshops which may be required from time to time. Initially you may be granted the authority to engage others to assist you in these areas. Of course, over time you will be expected to acquire or develop the capability to carry out all of these associated tasks.

**The other role...is the one which requires someone who can react quickly in what might be a very dynamic or chaotic situation.**

The other role which you will need to become familiar with is the one which requires someone who can react quickly in what might be a very dynamic or chaotic situation. You will need to be capable of organizing or leading the exodus whenever there is a serious threat to the health and safety of the population for which you are responsible. In this role you will need to be capable of an instantaneous response at any time; although, hopefully, you may never be called upon to exercise these skills in an emotional situation. One is inclined at first to think of a caged tiger who is always growling and eager to leap into action as soon as the opportunity arises. However, it is perhaps better to think of the apparently placid alligator that simply bides its time, reserving its energy until the moment when an opportunity for action presents itself.

Whoever is given or takes up the challenge of this role needs to be capable of thinking on his or her feet, working under pressure, keeping track of what is going on, giving clear orders, liaising with the emergency services, and ensuring that everyone is kept informed. It also helps if he or she can keep smiling and motivate the rest of the team.

## 1.6.1 Liaison with Other Services

To facilitate your evacuation and all the subsequent issues which might arise, you need to work closely with the emergency services while they occupy your site. You should also take into account the interests of the business continuity

(BC) team who may be anxious to organize or initiate various kinds of recovery and salvage operations.

The emergency services may wish to limit access to the site or to specific areas. In this case, someone may have to negotiate with them in order to have controlled access to key areas for special purposes such as retrieving backup. They may need to know more about specific hazards associated with the site or whether there are particularly valuable or delicate resources you would want them to protect. Your input could be very helpful to them in setting priorities for their work.

Almost certainly they will want to preserve evidence or carry out forensic work to establish the cause of the incident. You may be able to help them identify or locate particular items of interest.

They may have concerns about health and safety to the public, themselves, or your staff. Your intimate knowledge of the site, its resources and structures may help them in these matters. You should also consider the need for liaison with others who may be affected by the incident or its consequences such as neighboring organizations.

Where there is a fully functioning BC program in place, much of this liaison work will probably come under their responsibility, in which case you will need to work closely with them. It is also possible that the security officer will act as the liaison link. Whoever is given this responsibility will certainly appreciate your input, and you will no doubt want to know what is happening from that person's perspective so that you can understand and appreciate the situation together with the likely developments.

### 1.6.2 Single Point of Contact

You should establish a single point of contact for on-site liaison. The liaison officer or site controller should be clearly identifiable and in a prominent position. If, for some reason, that person needs to be somewhere out of sight, you will need to arrange for some temporary signs to indicate where the responsible person can be found. Once again, this responsibility may well lie with the security or BC team; however, even if one or both of these two groups exists, you need to be prepared for the possibility that an evacuation may not trigger a response from either one.

### 1.6.3 Communications

In an emergency situation there may be simply too much communication to deal with. There may also be conflicts between the natural desire and need to deal with incoming calls and the necessity to make outgoing calls. The event may also have seriously affected your communications capabilities.

Under these circumstances you need to be able to retain control and manage the limited resources at your disposal. It is time to apply these basic ground rules.

- Separate inbound from outbound calls. Restrict incoming calls to one number or a set of numbers that is only known to those who have a real need to call you. Make your outbound calls from a different number or set of numbers so there is always a line available for you to make urgent calls.
- In the same way, you should separate internal from external calls so that you have better control over the types of message and the information that is passing around. Staff can be kept up to date with some form of voicemail system. Others may need to be handled more cautiously or interactively.
- All enquiries from the press or any other casual enquirers should be referred to whoever is the nominated press or public relations officer, a person who should be properly prepared and trained for the role. Where there is someone available to speak on behalf of the organization, you should point out to them that you don't have the full facts and aren't in a position to deal with their questions but you can refer them to somebody who has the full picture. In the absence of such a person being available, you should politely inform any inquirers that your attention is focused on trying to ensure that everybody is safe and well.

### 1.6.4 Security

Any emergency situation is an obvious opportunity for criminal activity. Sometimes these shocking events are created as a diversion to occupy the attention of everyone while the instigators carry out their felony or misdemeanor. In any case, the petty criminal will often seek to take advantage of any situation in which there is a degree of confusion, unusual activity, strangers coming and going, and gaps in the normal security measures.

If the police are involved, they may provide some level of protection while they are on site, but your security is not their key concern. Once the incident is over and they have completed their investigations, they will want to return to their regular duties. The same is true of the fire and rescue service although they may want to stay a little longer while they ensure that there is no further risk of re-ignition or further outbreaks of fire.

You will need to ensure that an appropriate level of security is set up and maintained throughout the whole of the incident and the return to normal. This should be covered by those who are responsible for security on your site, but it is well worth finding out specifically who has that responsibility and whether they are properly prepared to deal with such a situation.

## 1.6.5 Access Control

Once the premises have been evacuated, you will need to make sure that everyone has managed to reach safety. This is an essential element of the evacuation procedure. From then on, a number of people may require access to the empty premises for a variety of reasons such as the retrieval of items or information of value, reconnaissance, surveys, or inspections. You, or someone such as the security officer, will need to be quite firm in order to obtain and retain control.

**It is essential that you, or whoever is responsible, should keep track of the number of people on site at any one time and who they are.**

It is essential that you, or whoever is responsible, should keep track of the number of people on site at any one time and who they are. You also need to know what they are doing and why they are doing it.

All those who have a valid reason to be on the site should be positively identified before they are allowed access. Before anyone is allowed onto the site that person must be able to provide satisfactory answers to the following questions.

- Who has asked him or her to be here?
- What is his or her objective and how will he or she achieve it?
- Does the person have the skills and does he or she need assistance, support, or permission to carry out the proposed task(s)?
- Why can't it wait until afterwards?

If a person requires support, assistance, or permission, the person should not be allowed on to the site until these matters have been resolved.

Keep opportunists and glory hunters out.

Key personnel such as the liaison officer, security officer, site controller, or person in charge of directly controlling the organization's resources at the site of the incident (called "bronze control" in the UK) should be easily recognizable. Such authorized personnel should wear reflective vests with labels indicating who they are. Their safety helmets should also have a similar label.

## 1.6.6 Inventory Control

The confusion that is the normal state of affairs during and after an emergency offers many opportunities for things to go missing. They can simply be mislaid or misplaced. If they are mislaid, someone needs to track them down. Misplaced items may be lost forever; they might become damaged or be cleared away with the rubbish.

As I have already pointed out, there is also the distinct danger of people making use of the opportunity to borrow or steal things. With so much movement going on, their nefarious activities are likely to go unnoticed. Additional, high profile, security should be introduced as early as possible.

## 1.7 Evacuation Triggers

To reinforce your argument about the need to take evacuation planning seriously, your bid for support should take account of the many and varied circumstances that might make it necessary to evacuate your premises.

Each of these scenarios may require interpretation, modification, or extension of the basic plans which you prepare. Some of them may warrant a distinct and different strategy to be considered and implemented where the environment has certain characteristics which might lead to, or exaggerate the impact of, a particular type of emergency.

Obviously, you will conduct a thorough investigation into the possibilities once you get the go-ahead to proceed to the next and subsequent phases. Your investigation should highlight a number of reasons why planning is necessary and give you a thorough understanding of what needs to be achieved. However, at the outset, you need to be aware of the range of possibilities because they may help you with your initial argument when bidding for support and funding. Later on, you will be turning some of these theoretical possibilities into realistic probabilities.

### 1.7.1 Classes of Incidents that Warrant Evacuation

- *Scale:* The scale of an emergency is the obvious geographical dimension, which may span anything from a part of your building up to, and even beyond, your whole country.
- *Speed:* Speed is the rate at which an emergency situation approaches and takes effect. This is a factor which you may need to take into account when developing your strategy.
- *Visibility:* Visibility, in this context, is the amount of warning which one might expect. For example, a hurricane warning may be posted several days before it is likely to impact your premises; this is an example of what I call an expected event. On the other hand you may have little or no warning in the case of a fire or an explosion – a sudden event.

While it is possible to estimate the degree of risk and the probability of the expected events in terms of scale and speed, it is virtually impossible to foresee the precise timing of any of the sudden events. Thus, you must be prepared to arrange an orderly evacuation in expectation of a known threat, perhaps at relatively short notice. You also need to be prepared to respond almost instantaneously to a sudden event which occurs without warning.

In the expected event, the pace of the evacuation will be relatively calm and casual, with time enough for everyone to get adjusted to the upcoming situation. In the case of a sudden emergency, however, everyone will be under pressure and time will be at a premium; success will depend to a large degree on the competence of those in charge of the evacuation. Such competence can only come about as the result of practical experience gained during tests, drills, and exercises.

**In the case of a sudden emergency...success will depend to a large degree on the competence of those in charge of the evacuation...the result of practical experience gained during tests, drills, and exercises.**

## 1.7.2 Six Emergencies Likely to Warrant an Evacuation

### *Group 1 – Contamination*

This group includes all those events where the environment is contaminated to the degree that it becomes hazardous to the health and safety of those in the affected area. Usually the cause of such incidents can be recognized in advance, and they will be included in the relevant risk registers. They tend to be creeping events which increase gradually in their impact; thus, one would normally expect to have some notice of the need to evacuate in a controlled manner. This notice might be a matter of hours or even days. As a general rule, it would seem that the larger the scale of the incident, the longer the notice might be. On the other hand, a large scale will usually imply an extended evacuation distance for the whole community, meaning that the emergency services may well take command of the mass evacuation of the whole population.

This group would include such events as oil spillages; chemical spillages; nuclear accidents, and atmospheric pollution.

### *Group 2 – Acts of Nature*

Acts of nature are those disturbances which nature throws at us related to climate, geography, and the seasons. In theory, we can reduce many of these risks through the choice of a suitable location. However, climate change and many other factors mean that most of us now work and live in parts of the world where natural events may threaten us from time to time. Some regions or zones are known for their susceptibility to particular types of natural hazard, and some times of the year are more likely to trigger weather-related events. Presumably, you are familiar with the natural hazards which are associated with your location. If not, you would be well advised to contact your local emergency planning officer, who will be able to provide you with information regarding the natural hazards which you might face.

Typically, acts of nature can be expected to affect relatively large areas, and some of the effects may need to be dealt with at a national or even international level. Your evacuation strategy must take such considerations into account.

This group would include such hazards as floods, earthquakes, volcanic activity, storms, blizzards, and lightning strikes.

While localized flooding may result from storms, it can also be caused by burst pipes. While a lightning strike can be expected to be an isolated incident, there are parts of the world where significant thunderstorms and lightning are prevalent at certain times of the year.

## Group 3 – Acts of Man

These disturbances are caused, knowingly or unknowingly, by people. Some of these are purely accidental; others are the end result of carelessness or ignorance, while an increasingly large number are pre-meditated attacks launched against individual organizations or a particular community.

This group would include such incidents as fires, explosions, terrorist attacks, disorder such as strikes or riots, and plain vandalism.

Generally, you can assess the risk of fires and explosions through exploring your local environment and the hazards which are known to exist. This research should lead to mitigation measures aimed at reducing or eliminating the threat. Acts of violence such as terrorist attacks, or acts of protest such as riots or strikes, are rather more difficult to assess or manage. However, the likelihood of such events can be monitored, and certain organizations or locations must be regarded as higher risk than others. Vandalism is also rather hard to predict although it must be recognized that it is much more prevalent in certain districts, and community policing combined with good security measures should reduce the likelihood.

All you can do is make sure you are prepared to evacuate all of your people safely at relatively short notice. Also, avoid complacency and remain alert to what is happening in the world around you at all times.

## Group 4 – Infrastructure

Infrastructure threats result from risks to what is commonly known as the "built environment" as opposed to the natural environment which existed prior to the industrial age. While it is all functioning properly, our modern infrastructure supports our community and makes it possible for us to thrive. However, when aspects of this infrastructure start to malfunction or fall apart we could be in considerable danger.

This group of threats to the infrastructure would include the loss or failure of services such as the power supply or the water supply. Either of these has a number of implications which might trigger the need to evacuate. The other type of infrastructure threat is the collapse of physical structures such as buildings, bridges, or elevated highways; these may be triggered by happenings both predictable and unpredictable.

### *Group 5 - Technical Failures*

While technical failures may cause a great deal of inconvenience, in most cases they are unlikely to initiate an evacuation. However, you should be aware of these areas of concern and consider the potential implications or impacts within your own particular environment.

This group of failures would include such problems as viruses and hackers, or failures in the hardware, software, supply chain, or production line.

Solutions and mitigation measures within this group of concerns would normally fall within the responsibility of the business continuity manager or the IT department.

### *Group 6 - People Problems*

The last group of problems which we should consider are the issues related to personnel which might affect our normal operations.

The two principal concerns here are a hostage situation or an active shooter. One should also think about an epidemic as a cause for concern, but, as a general rule, epidemics do not lead to an emergency evacuation; they require an orderly exit of the premises and, perhaps, isolation or treatment of those who may have been infected.

In the case of a hostage or a shooting incident, the emergency services will normally expect to organize and manage the ad hoc evacuation of the endangered area. However, you can make their job much easier if you have a tried and tested set of evacuation procedures which take people to a safe assembly area. Of course, it will make things much easier for all concerned if you liaise with the emergency service in advance of the actual incident. You can take advantage of their general knowledge and broad experience of the subject in exchange for your detailed site knowledge and proven procedures.

## 1.8 Coordination with Business Continuity Management

In today's competitive and relatively mature business environment, it is quite likely that your organization already has a business continuity management (BCM) program and business continuity plan (BCP) in place. Because of the many similarities and common ground between EEP and BCM, it makes a lot of sense to coordinate the two functions. Possibly, working together could be achieved by combining the responsibilities into one role or through setting up a center of competence which embraces both disciplines.

**The lifecycle, the aims, and the products of an EEP are quite similar to that of a BCP...they are aimed at the same target audience, and the outcomes from each stage are more or less variations on the same theme.**

If the company does not already have a BCM program in place, then it is quite likely some managers are contemplating the possibility of acquiring the benefits of assured survival that such a program can bring in a world of increasing uncertainty, intense competition, and rising expectations of the consumer population. This sort of awareness would provide you with a golden opportunity to link up with the proponents of such a program, generating a great deal of synergy by co-coordinating these two endeavors. Rather more valuable outcomes from a business perspective will be the increases in efficiency and effectiveness which can be gained from such a move.

As we have already seen, the lifecycle, the aims, and the products of an EEP are quite similar to that of a BCP. Indeed, the lifecycle which I am proposing here within this book is derived directly from the standard, generally accepted BCM lifecycle that is the basis of BS 25999 (British Standards Institution, *BS 2599-1:2006*, 2006) and of the *Good Practice Guidelines* (2010) of the Business Continuity Institute (BCI), the worldwide professional body for BC practitioners.

The principal benefits of amalgamation arise from the fact that most of the procedures within the two disciplines are similar in nature, they are aimed at the same target audience, and the outcomes from each stage are more or less variations on the same theme. Therefore, it makes sense to combine the two parallel activities into one rather more comprehensive concurrent activity, making the overall process far less intrusive and much more efficient. Also, it will result in a fully integrated set of outcomes such as plans, procedures, facilities, and resource utilization.

It is worth bearing in mind that for most BC professionals, BC activities are neither their original nor their only responsibility. The vast majority of practitioners come to the subject by accident or inspiration after gaining experience in some other, probably unrelated, field of endeavor. This varied experience means:

- They usually bring a broad perspective to their work and are quite capable of thinking outside of the box.

- They are normally quite innovative by nature.
- Their practical approach could prove to be very useful.

If you do manage to link up with them, I am certain that they will welcome your attention and offer you their full support when you discover them and approach them.

However, before you set off with high hopes and a burning ambition to unite these two disciplines, it is worth pausing to consider comparing their aims, methods, and subtleties. If you are going to reap the full benefit, you must avoid the obvious pitfalls, appreciate the differences, and recognize which aspect you should be focusing on at any one time.

## 1.8.1 Similarities and Differences Between BCM and EEP

Both of these disciplines have protection and assurance as their primary aims, and both use the principle of thinking ahead as the basis of what can and should be achieved. Also, they both use risk assessment as a key tool in developing their strategy and tactics; however the EEP emphasis is on the health, safety, and welfare of the people, whereas BCM is more concerned with functionality, availability, and access to resources and facilities.

Another common feature is that both involve plans that rely on the way people actually carry out the proposed actions or activities. Thus, both BCM and EEP plans need to be tested and exercised properly in order to be effective.

BCM and EEP share the need to be prepared and able to cooperate and liaise with the emergency services in the wake of an incident. They also tend to form close links with the security people and the preventive measures which they put in place, especially in connection with an emergency response situation.

## 1.8.2 Key Differences

Here is a list of some of the key differences between the EEP and BCM approaches in terms of thinking, attitudes and expectations. There may be others which you might encounter.

| Emergency Evacuation Planning | Business Continuity Management |
|---|---|
| Requires people to conform to the plans and follow instructions. | Encourages people to interpret the plans and be innovative. |
| Is entirely people focused, i.e., aims to ensure the safety and welfare of all people. | Is principally business focused, i.e., aims to ensure survival of our business. |
| Is mostly about the usage of safe and secure pathways to identified assembly areas, i.e., relatively low-tech concerns. | Is mostly about the availability of, and access to, premises, data, and technology, i.e., relatively high-tech concerns. |
| Can be seen as a legal requirement or associated with a legal requirement. | Is purely voluntary in most industry sectors. (Mandatory in the financial sector.) |
| Normally a standalone activity which doesn't involve any third party participation. | Often involves the support or assistance of suppliers, service providers, and regulators. |
| Customers and suppliers are usually unaware and unconcerned. | May be a contractual requirement within the supply chain. |

### 1.8.3 Common Ground

In both the EEP and the BCM version of the lifecycle model, there is a constant need to promote awareness and training. Because of the similarity and links between two subject areas, it makes sense to combine these parallel activities into a single comprehensive awareness and training program under a common heading such as "Protection and Preparedness." Each of the two disciplines will benefit from a higher profile as a result of the increased area of overall responsibility and wider impact.

There is no doubt that combined evacuation and incident response exercises will be much more cost-effective than trying to run two similar but different events. It will also be less confusing for those involved in, and those who hear about, these proceedings.

Wherever there is a mature BC program in place, there is an ongoing need for review and maintenance work to be carried out on a regular basis. This is often achieved through the appointment of local part-time administrators who are trained and authorized to act as the local representatives for the subject. Within their section or area of activity, these BC administrators are expected to keep the plans and procedures up to date, ensure any special resources are retained and maintained, update their section of the risk register, and promote awareness and interest in the subject. This approach normally includes a central center of competence which these administrators can use as a source of reference and repository for documents, resources, and intelligence. Clearly, these BC administrators could also become useful agents for the EEP program if they were given a little training and some encouragement. They might act, or recommend someone who should act, as evacuation marshals or assistants in the event of an emergency. Apart from that, they could prove to be very helpful to those people in their section or area who need to develop their own PEEPs. Using their local knowledge and accessibility, they could work with those concerned to ensure, or guide, the development of customized plans which take full account of the individuals' needs and the intricacies and nuances of their surroundings.

## 1.9 Obligations and Responsibilities

Although no commonly accepted international standard for emergency evacuation procedures or signage has been established, obligations remain for employers and property owners to afford a certain degree of safety in the workplace. Generally these obligations are outlined in rules and regulations which impose, refer to, or describe standards in respect to plans and signage regarding emergency evacuation, fire protection, and fire prevention.

**Corporate citizenship and social responsibility should be the driving force behind making the workplace, dwelling, or pleasure area safe for occupants and visitors alike.**

The rules and their observance may vary from one jurisdiction to another. Sometimes they are rigidly enforced and at other times and in other places the regime may be much more tolerant or even lax. The authorities always seem to be more observant and conscientious in these matters in the wake of a major emergency or disaster although the risks do not increase simply because of one well known or publicized local incident. We need to be properly prepared and trained at all times. Disasters do not seem to recognize high and low seasons; they just happen, although there may be a tendency for particular types of events to occur at certain times of the year. However, there is enough variety and uncertainty to give us cause for concern at all times, enough to justify the precautions which are being promoted here. Corporate citizenship and social responsibility should be the driving force behind making the workplace, dwelling, or pleasure area safe for occupants and visitors alike.

## 1.9.1 Legal Obligations

In most modern jurisdictions, employers and those who are responsible for the management of facilities where people gather for business, pleasure, health, or learning have a legal obligation to conduct risk assessments; to use health and safety signs in areas which may contain hazards; and to ensure that all fire escape routes are properly marked. The details of the relevant legislation may vary, but the principles will remain the same throughout the world.

For the moment you only need to be aware of the fact that there are a host of rules, regulations and standards which you may need to take into account during the development and delivery of your EEP plans and procedures.

You must also appreciate that the weight of the law is behind the need to ensure the health, welfare, and safety of all those who use or visit your premises.

# Phase 1 – Key Actions

- **Do your homework.** Understand national and local building regulations, national and local fire regulations, and the infrastructure of your organization.
- **Gain sponsorship.** Your sponsor needs to be someone in authority, who has the confidence of top management, and will support you in taking responsibility for the health and safety of everybody in an emergency.
- **Establish the policy and scope for the EEP program.** The policy will be your benchmark and source of authority throughout the program.
- **Prepare a business case to justify the effort and expenditure.** Legal and moral implications should be made clear by referencing specific laws, rules, and regulations which apply to your region, culture, and industry sector.
- **Establish a management structure and protocol for the EEP program**. Indicate clearly who is responsible at each level of management, decision-making, and administration.
- **Agree upon an overall EEP strategy for your organization**. Stay flexible to adapt the strategy to suit different areas and populations.
- **Develop the tactics for each area of the organization.** Take account of any significant variations within the population, such as age, ability, and any special needs.
- **Set up liaison with local emergency services.** Where there are a number of different locations, appoint a liaison officer for each of them.
- **Create a robust communication procedure**. If the available resources are inadequate, do whatever it takes to acquire suitable equipment.
- **Establish suitable security measures.** During the response to the incident and the subsequent recovery and restoration, put in place security measures for access control and inventory control.
- **Coordinate with existing business continuity programs**. If there is no business continuity operation in existence, plan ahead for EEP to complement BC in the future.

## Discussion Questions – Phase 1

At this point, you should have a good overall grasp of what EEP is and could be. You should also be starting to realize what it might mean or become within your own environment.

1. Think about the times you have been in an office, hotel, or business that needed to be evacuated – either as a drill or because of an actual emergency. Was there any confusion? What worked well and what could have been handled better?
2. Who do you think should "own" and who does "own" EEP within your organization? Is ownership formally determined and documented or is it simply assumed? Does it matter and if so why? Would you want to alter or improve the situation? How would you approach implementing any changes in this area?
3. We are using our EEP Lifecycle model as the basis for an important program of work. Do you find that such a structured approach helps you to understand the subject? Are you familiar with any other lifecycle or program model which might be adapted to create and implement an EEP?
4. We propose a five tiered administration structure which runs from policy down to instructions or signs. Does this fit in with what you would regard as normal practice in your environment? How would you adapt the structure and terminology to your organization and culture?
5. In the text we referred to the Plan-Do-Check-Act (PDCA) model for managing a system. How would you explain to your manager and co-workers that EEP should be seen as an ongoing management system rather than a one-time project? How would the PDCA model fit in the way your organization traditionally manages or monitors systems which you will be encouraged to use?
6. Which of the "evacuation triggers" we describe are most likely to occur in your work environment? What points would you make in favor of better preparation for a specific anticipated emergency (e.g., fire, flood, etc.)? What options would you suggest?
7. What are the key areas in which EEP and BCM could quite naturally coordinate their efforts? If you were making a presentation to a manager in charge of BCM, what would be some of the advantages to BCM that you would give as arguments in favor of this type of coordination?

# Phase 2

## Embed EEP Into An Aware and Prepared Corporate Culture

This second phase of the emergency planning lifecycle is all about raising awareness and embedding the subject into the establishment's culture, which is perhaps the most nebulous aspect of what you are trying to do because there are no obvious tangible products from this part of the program. The principal change you are trying to make here, as you embed emergency evacuation planning (EEP) into the culture, is to alter attitudes and behavior which happens only within people's minds.

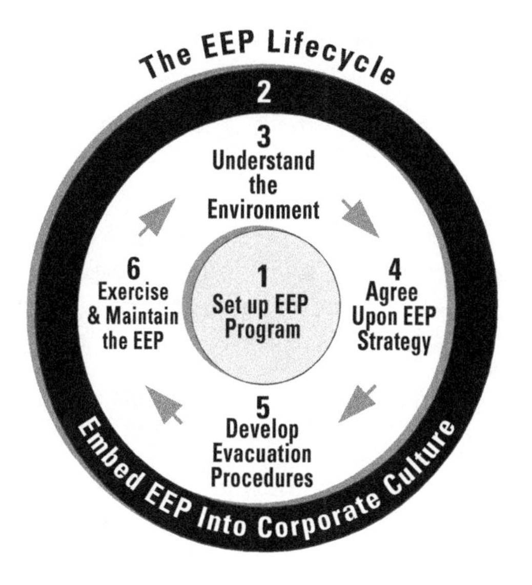


*This section will help you to:*

- ➢ Relate the BCI Good Practice Guidelines to your EEP effort.
- ➢ Create an effective training and awareness program.
- ➢ Publicize EEP within your organization.
- ➢ Coordinate the EEP program with BCM planning and activities.

## 2.1 BCI Good Practice Guidelines

The EEP lifecycle model in this book has its roots in the Good Practices Guidelines of the BCI, which can be adapted to EEP. Of course, you recognize that business continuity (BC) is a rather more complex subject with innumerable side issues; therefore, BC planners may need to communicate a slightly more involved set of messages. On the other hand, your EEP messages are equally valuable and are important to every single soul who ever enters your premises.

**Awareness, training, and communication are the main tools to achieve this recognition within the corporate culture...**

The BCI notes that having a proper management system in place brings a number of advantages both to the subject area and to the organization as a whole. The same principle applies in the case of EEP, although the specific advantages are slightly different due to the nature of our cause. The benefits of a fully functioning EEP program include ensuring that an organization can:

- Manage an emergency evacuation much more efficiently.
- Instill confidence in its stakeholders, especially staff and customers, in its ability to deal with an emergency.
- Enhance its emergency response and human welfare capability over time by including EEP implications in strategic and tactical decisions at all levels.
- Minimize the impact and likelihood of emergencies.

These benefits are likely to be fully realized only if the culture of the organization understands and appreciates the need for EEP and actively promotes its growth across the organization. Therefore, you should ensure that:

- The EEP program is set up in a manner fit for its purpose.
- Confidence is instilled in stakeholders, particularly staff and customers, that anyone who is on the premises at the time of an incident would manage to reach safety without being subjected to trauma or unnecessary delay.
- Probability of such incidents will be reduced and (if they should occur) their impact on those present and on the business operation will be minimized.

Awareness, training, and communication are the main tools to achieve this recognition within the corporate culture; these tools need to be used on a regular and ongoing basis. Throughout the program you have to consider the "What's in it for me?" factor. Without a perceived personal benefit, most people tend to ignore or forget much of the information with which they are bombarded daily in the world of mass communication. Other phases of the planning lifecycle have an identifiable beginning, middle, and end. In contrast, maintaining awareness, as part of the organization's culture, will be an ongoing project.

**One of the early messages or lessons will be making everyone aware of the new EEP policy (from Phase 1 of this process) and its implications.**

### 2.1.1 What is the message?

A natural question might be: If we don't yet have an EEP, what is the content of the "training" at this point in the process? That's a good question. At the beginning, you will be talking about work in progress; the emphasis will be on awareness and a review of current practices and policies. The company may have in place some emergency procedures and supplies, sufficient to comply with minimum legal requirements. Procedures and supplies may include regular fire drills, some fire extinguishers, some first aid supplies, and fire exits with appropriate signage. Therefore, of necessity, your initial messages will be simple and contain a limited amount of information. Your purpose will be to announce your presence and make your audience aware of the benefits to the workforce of your policy and plans. One of the early messages or lessons will be making everyone aware of the new EEP policy (from Phase 1 of this process) and its implications. Remember that this training will consist of a series of lessons. As in any other training course, you will start out with basic information and build upon their knowledge gradually. In this case, you will also be learning as the program grows and develops.

Gradually, as you move forward with your investigations, you will have more concrete information to communicate to your audience. During this initial awareness phase, you will begin the education process with some basic knowledge and a few facts. As you progress through subsequent phases you will be building up from that knowledge base and adding to it in order to ensure that certain emergency behaviors become an integral part of the culture.

### 2.1.2 Training as an Ongoing Requirement

Training will be required as part of the ongoing requirements for employees. Thus, it will change and gain more detail as your project proceeds. At key points during the subsequent phases of the methodology, you will be adding further elements to the awareness program. At first it may be more theoretical, but as you develop and deliver your strategies, procedures, and evacuation plans, the emphasis will become more practical as you move towards establishing an ongoing routine in which the training becomes part of a regular exercise and maintenance program.

## *Tragic Outcomes can be Compounded by Insufficient Training*

### Costa Concordia

The tragedy of the Italian cruise ship Costa Concordia (operated by Costa Cruises, an Italian subsidiary of US-based Carnival) is an unfortunate example of an emergency evacuation that was handled poorly.

On 13 January 2012, the 114,500-ton ship struck a rock in shallow water just off the shore of the small island of Giglio on the Tuscany coast in Italy. In a freak accident, the ship rolled over onto her side in shallow water causing a completely unexpected emergency situation. The result was a type of emergency evacuation which nobody had considered or planned for: horizontal corridors to safety suddenly became near vertical obstructions, causing panic to set in among the crew and passengers.

The perception of passengers was that the situation on board the ship was chaotic, that they were plunged into darkness, and that the orders to abandon ship were delayed unnecessarily.

Investigators are seeking an explanation for the 68-minutes that elapsed between the time the ship hit the rock and the order was issued to abandon ship. Was it because the captain underplayed or underestimated the gravity of the damage sustained? Or was it reluctance on the part of the cruise ship owners to evacuate the passengers?

Stories from survivors have cast doubt on how well the crew had been instructed and drilled in carrying out the necessary steps for abandoning ship. Passengers have questioned whether members of the crew actually knew what they were doing, noting that the crew gave conflicting orders – some told passengers to go back to their cabins, while others told them to go to the lifeboats. Passengers related that they felt abandoned by those who were responsible for their safety.

Survivors also reported instances of an insufficient number of life jackets and of some lifeboats that were out of order and could not be launched.

Eleven passengers and crew are known to have died, and 21 persons were never accounted for. The rescue efforts were suspended at the end of January. The media's focus was entirely on what went wrong – as related in the hair-raising stories of survivors – rather than the fact that over 4,000 people managed to reach safety. However, as the investigation continues, the lessons from this disaster will include exploring how that loss of life could have been prevented and how the evacuation of the survivors could have been handled more smoothly and effectively.

## 2.2 Developing the Training Program

Before you embark on any sort of training program, you have to establish just what this program is supposed to achieve. You need to put into practice the same plan-do-check-act model that we outlined in Phase 1 – although in this case your starting point should be the check stage. You need to establish what your audience members already know about the subject and compare this with what you have determined that they need to know. This is what is known in the trade as a training needs analysis.

**Plan:** The *plan* stage of this training program is the point at which you figure how, when, and where you are going to deliver the training. If your organization has some sort of regular training program in place already, then you should tap into that and use the resources, tools, and skills of the existing training to plan and deliver your EEP training and awareness program. On the other hand, you may need to start from scratch, which means you have to become embroiled in the planning aspect of this program.

It doesn't need to be a massive research project, but you will need to ask a number of those to be trained what they know about the subject, what they think they ought to know, and how they think this knowledge level might be achieved. I always think it is a smart move to get those to be trained to help design the course material and develop the objectives. This participation ensures you get their buy-in and also means the training will cover the subject from their perspective and at their pace. Of course, you have to retain a degree of control and make sure that the outcome is a sensible and rational training plan; sometimes you have to take some of their comments or suggestions with a pinch of salt, especially where there may be hidden motives or a misunderstanding about some aspect of the subject.

There are two main areas of consideration in this training program.

1. For most, if not all, of the people who spend time on the premises in question, the principal concern will be getting the message across that specific emergency evacuation plans are in place, or under development. This audience needs to know the detail of the basic procedures, be reminded from periodically of the planned exit and escape routes, and be instructed about what is expected of them when they eventually reach the safe assembly areas.
2. The second area of consideration when putting together an EEP training program will be addressed to those who have special needs or interests beyond that of the population at large. Here I am thinking of those who will need to play an active role in the evacuation procedures or anyone who may need or want to be involved in the development or customizing of plans to suit particular individuals, groups, or areas.

**Do:** In order to achieve either or both of these aims, you will obviously need to choose a way to create a productive dialog which suits both the culture and the population concerned. This dialog may be achieved through formal training sessions or workshops, through the intranet, or in face-to-face conversations. It is probably best that you liaise with the human resources or personnel department because training and communication with the workforce are tasks for which they have overall responsibility. They may even have someone dealing specifically with the development and delivery of training courses and programs who can prove useful in helping you develop and deliver your messages in the most efficient and effective ways.

For the longer term, it might be useful to link up your awareness program with HR's existing orientation procedures which familiarize new employees with the ways and means of working within the organization. Sometimes the orientation process is supplemented or replaced by a staff handbook, in which case a page or so on EEPs and procedures would be a great help towards developing the requisite awareness and confidence which you are striving for.

**Check:** When you have completed your survey, you must collate and verify the data which you have gathered before analyzing it to determine what is needed and wanted in the way of training and the dissemination of information regarding EEP.

Awareness training and publicity are aimed at informing and consequently embedding the subject within the culture of the organization. This embedding in the culture is important because, according to the BCI, it is only when the culture is appropriate that an effective strategy can be implemented. It is here that you are trying to develop the backdrop for the action which is due to unfold upon the EEP stage. Later on, you will be asking the actors to take part in rehearsals or exercises which will give them the practical experience to develop the necessary confidence and competence in the plans and procedures.

**Act:** Once you have completed the check stage of the training program it is time to move on to the act stage where you work on how to correct any deficiencies which you discovered and reinforce the good qualities which you came across. In almost any organization, you can identify someone already responsible for managing, organizing, and possibly delivering in-house training, probably someone within the HR or personnel department. You and that person will need to agree who will arrange and deliver this program. Because EEP is a relatively straightforward subject, the instruction will not depend upon the use of high-level education skills; anyone who is a good communicator should be able to deliver this type of training. You will probably be given the opportunity to prepare and deliver these lessons yourself, perhaps with a little assistance at first, but you will soon get the hang of it.

The act stage is really about designing the content and developing the material which will be required to transfer the knowledge and develop the skills to enable the people to participate in your EEP program. The majority will only need to be given enough information and confidence for them to take part in a successful evacuation. Others may need your assistance to be able to come to grips with customizing or tailoring plans to meet the needs of particular areas, departments, or individuals.

The style of teaching will depend partly on the type of audience you have to face. Other factors will include your own personal preferences, what is easily available in the way of teaching aids, and the level of detail you feel you need to convey. The amount of detail will be governed largely by the scale and complexity of the site or location you are dealing with. In most cases, a flip chart and some colored pens will enable you to get your messages across.

The other common alternative to a flip chart is to use PowerPoint or some similar PC-based graphics tool which can convey words, diagrams, maps, and photos in a professional manner. Using a PC-based tool is a particularly useful approach, especially where you may have to run several similar training sessions. You can also mix and match a selection of the slides to suit your presentation to audiences with different needs or interests.

Before embarking on delivering an actual lesson to a class of strangers or colleagues, it is a good idea to have a couple of rehearsals to make sure you have your story right and to get an idea of the timing for the whole thing. You should also develop an outline script to make sure that your talk covers all the relevant points, gets them in the right sequence, and uses the right sort of language. Preparing a set of slides accompanied by a script, or a set of presenter's notes, would also mean that others could deliver the same presentation on your behalf, an advantage in a large organization spread over diverse sites.

You should also think about the value of providing the audience with some prepared notes such as an exit route and escape path map. This is a judgment call based on the complexity of the layout of the premises and what other support materials you plan to make available at some later stage, if any.

## 2.3 Presenting the Training

Once the training curriculum and the supporting materials – such as slides, presenter notes, and handouts – have been developed, it is time to work out how you are going to bring the key components of the training together. I have identified eight components which have to coincide in order to deliver an effective training session. These are:

1. Venue – somewhere suitable to run a session without disturbance or interruption.
2. Participants – persons identified as those who need or will benefit from the training.
3. Diaries or schedule calendars – ensuring that venue, presenter, and participants are available to attend.
4. Facilities – the tools, equipment, and resources required to present the training.
5. Materials – documents and reference materials to support the delivery of the training.
6. Presenter – someone who is capable and prepared to deliver the training.
7. Catering – drinks and snacks for all those attending.
8. Protocol – rules of the classroom to ensure the training runs smoothly.

The first six are essential; if you cannot bring the people and the equipment together on an agreed date at a specific place, then training simply cannot take place.

**Venue:** The preferred venue would be a meeting room or training room which is close to their normal place of work or residence. However, if there is no suitable in-house facility available, it may be necessary for you to rent a room in a nearby hotel or training center. In this case you have to approach top management for budgetary approval before finalizing the arrangements.

**All people with special needs should either be attending the training or be represented by someone who will be working with them in regard to EEP.**

**Participants:** As far as participants are concerned, your ideal target audience is 100% of the population, but in reality this may not be achievable. What is certain, however, is that all people with special needs should either attend the training or be represented by someone who will be working with them in regard to EEP. Others who should attend would include anyone who is likely to be appointed to a specific role, such as marshaling or assisting during an evacuation. In fact, it is preferable that all the marshals attend all of the sessions to show their support and commitment to the program. They would also be able help you deal with questions and answers at the end of the session; their local knowledge could be quite useful.

**Diaries or Scheduling:** Matching diaries or scheduling calendars to get a good representative attendance at each of the training sessions can be quite a challenge, especially in a dynamic organization in which everybody has a heavy workload. Since a dynamic enterprise and a heavy workload seem to be the norm these days, I usually find it takes three or four sessions to be able to capture the majority of the target population; realistically, I never expect to get 100% coverage. In the case of very large buildings or sites, it could take as many as seven or eight sessions to accommodate everybody, perhaps spread over a period of weeks. Of course, you would hope to have attendance at these training or education sessions made compulsory, but usually you will have to settle for voluntary participation, which works fine because most people do regard this as a subject worthy of their attention.

**Facilities:** When it comes to the question of facilities, I think that will depend upon the style of presentation which you are planning to use. These days I find that nearly all meeting rooms have adequate facilities for most types of presentation, but it is always wise to check what is going to be there on the occasion when you are expecting to use the facilities. If you are not familiar with the venue, you should have a practice run beforehand; try hooking up your laptop with their projector if that is how you will be presenting. Of course, if you will be taking your own projector and laptop with you, then you might take their compatibility for granted.

**Material:** Unless this training is merely a simple introduction to the subject, and the environment is pretty straightforward, you should have some handouts ready for the participants to take away with them. These handouts should be distributed at the beginning of the session so that the attendees can make notes on them if they wish. Since attendees need to be able to write in comfort, this suggests that the venue should be laid out in classroom or boardroom style so the people sit at a table or a desk. Make sure to have a few spare copies of the handout so that people can take away an extra one for any of their colleagues who weren't able to attend.

As for other key materials, we have already spoken about pens and a flip chart to make notes even if you don't intend to use the flip chart as a teaching aid. The alternative is to take a note pad and pen with you just in case you need to write down something useful or essential. I usually take a clipboard with me whenever I think there may be the chance of having to record something like a request or a learning point worth capturing.

**Presenter:** A good presenter is one who is enthusiastic and knowledgeable about the subject and is able to articulate that enthusiasm and knowledge in a manner which connects with the audience. If you are not passionate about the subject, then it is unlikely that you would want to be teaching others about it. Knowledge can be acquired, and by the time you have finished reading this book, you should be sufficiently well equipped to convince your participants that they should take note of what you are saying.

**Catering:** Catering is a matter of degree and will depend to a large extent upon the culture and the venue. If no catering of any sort is to be provided, then the participants must be advised of this in advance so they can make their own arrangements with regard to refreshments.

**Protocol:** Protocol is all about classroom behavior, and in most environments this can probably be left to the good manners and good will of the people concerned. However, in many instances I have found it necessary to point out the house rules either before or during a training session. Points to be raised or considered will include the use, banning, or control of mobile phones; incoming messages; timekeeping; religious or cultural considerations; toilet breaks; refreshment breaks; smoking; swearing; question handling; study periods; breakout sessions; classroom layout; and air-conditioning settings. You should also point out the fire alarm and escape procedures before starting the training. It may also be necessary to keep a register of attendees as evidence of their participation and to track those who have been missed out.

**If you do not show proper respect for their traditions and work ethics, it is likely that much of their attention will be focused on your bad manners rather than the subject at hand.**

With regard to religious or cultural considerations, experience has taught me to respect other people's religious or cultural beliefs and customs. Wherever I go, I make a habit of discreetly inquiring about any special considerations which I should take into account when planning or running a training program. If you are already familiar with the culture and the people you will be working with, then I expect you will automatically adopt an acceptable stance in this regard. If you do not show proper respect for their traditions and work ethics, it is likely that much of their attention will be focused on your bad manners rather than the subject at hand. They are also likely to reject your ideas and beliefs in the same way as you seem to be ignoring theirs. Ignorance is not a valid excuse although their tolerance may lead you to think otherwise.

## 2.4 Publicizing Within the Company

In order to complete the EEP backdrop, or set the scene, for successfully acting out the scripts of your EEPs, you will need to publicize your intentions and achievements throughout the company using whatever means are at your disposal. You must aim to capture the broadest audience possible which requires you to keep the message simple, repeat it often, and spread it as wide as you can.

We will look at a few examples of organizations that have taken approaches which suited their needs, their audience, and their culture, but you can explore many other ways to get your message across to the target audience.

Recently, a client and I came up with the idea of a handout, or broadsheet, outlining the basic principles of EEP as an introduction to the subject which we could distribute as a kind of starter pack to kindle interest and set people out on the right path towards developing the cultural side of our emergency evacuation program.

What we were looking for was a set of guidelines based upon a series of milestones that described in the very broadest terms how people should think and behave in response to an emergency announcement.

Our first thoughts resulted in a 10-step generic outline of how we expected people to behave in response to an alert or an alarm. We summarized this as a single page for use as a handout or a poster. We also approached the personnel department and arranged for this page to be added to the staff handbook. An image of the handout version is shown here.

## A Business Company

### Generic Emergency Evacuation Guidelines

*These guidelines provide an overview of the generic emergency evacuation procedures for all those who visit, work, or reside in our buildings. There will be specific plans for each particular building giving details of the local exit routes, exit points and escape paths which lead to the designated safe assembly areas.*

*In an emergency situation, you should consider those who visit, work or reside, close to you; watching out for each other in pursuit of mutual benefit and moral support.*

1. **Alert or Alarm**
   a. Alert - When the fire alarm sounds intermittently, a possible danger is being investigated and you should remain alert but do not start any new tasks.

   b. Alarm - When the fire alarm sounds continuously there is a real danger and you should prepare to leave the building without delay.

2. **Listen**
   - Whenever there is an 'Alert' you should listen for an announcement giving directions regarding the escape route to be used

3. **Directions**
   - Take heed of any directions which may be given and ensure that you and your immediate colleagues (your group) heard and understood the message.

4. **Close**
   - Close down, take your handbag, grab your coat and prepare to leave. Close the doors behind you if you are the last to leave.

5. **Proceed**
   - Make for the nearest fire exit.
   - Follow the signs (unless directed otherwise).
6. **Exit**
   - Leave the building via the designated fire exit.
7. **Escape**
   - Take the escape route toward the designated assembly area. Follow the signs.
8. **Assemble**
   - Gather at the assembly area and check that all of your group have arrived safely.
9. **Report**
   - Ensure that you all report to the marshal who will be at the marshalling point wearing a high visibility jacket.
10. **Wait**
    - Remain at the assembly area and await further instructions.

A template version of this document, "Generic Emergency Evacuation Guidelines," is included in the EEP Toolkit for you to adopt and modify to suit the requirements of your own organization and the environment in which you operate.

## NYC Hazards: Hurricane Evacuation

Under New York State law, the Mayor has the power to declare a local state of emergency. This might include issuing evacuation instructions for one or more ***hurricane evacuation zones*** if it were determined that clear and present danger to the public exists.

Deciding to issue evacuation instructions requires in-depth analysis of ***storm forecasts*** and local conditions, which is coordinated by the Mayor, OEM, State and Federal agencies, the National Weather Service and National Hurricane Center, and jurisdictions throughout New Jersey, Long Island and upstate New York.

***Find out if you live in a hurricane evacuation zone***

## Evacuation Instructions

The Mayor can issue two different kinds of evacuation instructions:

***EVACUATION RECOMMENDATION***: The Mayor may recommend certain residents take steps to evacuate voluntarily. A recommendation might be issued to cover residents of certain zones, communities or building types. An evacuation recommendation could also be issued for the benefit of people with mobility challenges who need extra time to evacuate.

***EVACUATION ORDER*:** The Mayor may order residents of specified zones or communities to leave their homes for the protection of their health and welfare in the event of an approaching storm.

## How to Evacuate

Since flooding and high winds can occur many hours before a hurricane makes landfall, it is critical evacuees leave their homes immediately if instructed to do so by emergency officials. Evacuees are encouraged to seek shelter with friends or family or outside evacuation zones when possible.

To avoid being trapped by flooded roads, washed-out bridges or disruptions to mass transportation, evacuees should plan their mode of transportation with special care.

- **Plan to use mass transit as much as possible**, as it offers the fastest way to reach your destination. Using mass transit reduces the volume of evacuees on the roadways, reducing the risk of dangerous and time-consuming traffic delays.
- **Listen carefully to your local news media**, which will broadcast reports about weather and transportation conditions.
- **Evacuations from at-risk zones will be phased** to encourage residents in coastal areas to leave their homes before inland residents and to help ensure an orderly evacuation process.
- **Leave early.** Evacuations will need to be completed before winds and flooding become a threat, because wind and heavy rain could force the early closure of key transportation routes, like bridges and tunnels.

The City advises against car travel during an evacuation. The City will be working hard to keep roads clear, but traffic is unavoidable in any evacuation. Driving will increase your risk of becoming stranded on a roadway during an evacuation.

## IF YOU MUST TAKE A CAR:

- Be ready for a long, slow trip. Be aware the City will deploy public safety personnel along major transportation routes to help vehicular traffic flow as smoothly as possible. Have a full gas tank before you go.

- Stay tuned to local media for information about road and bridge closures. New York State's 511 can help you monitor traffic on State roads.
- Evacuation Centers are the ONLY places where people may park vehicles. Many evacuation centers do NOT have parking available. Tune in to local media for instructions.
- Large vehicles may be prohibited in windy conditions. This could apply to trailers, trucks, boats and other vehicles with a higher wind profile than a car or SUV.
- In any significant rainstorm, **avoid driving through standing water** if you cannot tell how deep it is.

If you must go to an ***evacuation center***, it is important to carefully select what you take with you. Do not bring more than you can carry, but be sure to bring your Go Bag with you.

*— Source: New York City Office of Emergency Management (OEM), http://www.nyc.gov/html/oem/html/hazards/storms_hurricaneevac.shtml, 24 Jan. 2012*

## Finding the Right Catchphrase

Your plan will benefit from creating a memorable catchphrase that, essentially in shorthand, conveys the essence of your emergency message. Many of us as children learned a basic phrase for taking care when we were in the vicinity of railroad tracks or railroad crossings: Stop, Look, and Listen. It was short, action-oriented, easy for a child to remember, and undoubtedly saved many lives. Recently, Amtrak in the U.S. is using a memorable security phrase to encourage passengers to report any suspicious behavior or items on the trains or in the stations: *If You See Something, Say Something!*

Largely as a result of what happened on 11 September 2001, the UK government recognized the need for London to be able to respond quickly and effectively if a similar incident occurred in the capital. A coalition of key agencies (known as the London Resilience Partnership) was brought together by the deputy prime minister in May, 2002, to plan and prepare for potential emergencies. This was the first time a strategic, pan-London regime had been established to coordinate planning across London. The coalition brought in a wide range of subject matter experts to advise and guide the group in their deliberations. I participated in that group, which developed guidelines, advice, and tools for use by local authorities, businesses, and people.

London Resilience came up with the catchphrase *Go in, Stay in, Tune in,* advising people to remain indoors and listen to the local radio in the event of a major emergency. The beauty of this simple three-part phrasing was that it had rhythm as well as being clear and easy to remember. It made good sense and, as a result, it was adopted by all of the agencies. This advice soon became an integral part of the emergency planning culture in my part of the world.

In any organization, when a similar phrase is used, it is designed to act only as a reminder to reinforce the details of evacuation as described thoroughly in the guidelines. However, I do believe that simply creating, publicizing, and reviewing the catch phrase is a beneficial exercise that could help people remember what to do in an emergency. At least they have spent some time thinking, and therefore learning, about the subject.

When I visited the Chartered Institute of Environmental Health (CIEH), I was pleased to notice that available in the reception area for arriving visitors were little cards about the size of a post-it note which carried brief instructions about how to escape from the building and where to go in the event of an emergency.

## Evacuation Stories from Employees as Part of Publicizing the Program

Many people have unusual or unexpected fears or disabilities which they tend to keep hidden, and EEP does need to take these idiosyncrasies into account. Sometimes these concerns are of relatively minor importance, but they could also be quite significant. The point is we don't know until we ask or have to find out the hard way.

- One organization, as part of publicizing its EEP program in a rather different way, asked staff to submit any little "war stories" which they might be willing to share regarding previous experiences or their thoughts about getting safely out of the building. These anecdotes were then published in the house magazine and readers were invited to vote for what they considered to be the best story. The personnel manager promised to take the winner out to lunch.
- One of the entries, submitted by someone who was rather slow on his feet due to osteoarthritis, told about an occasion in his previous employment when employees had been forced to flee the building because of a fire. As he struggled down the stairs, he felt slightly embarrassed, concerned that his slow progress was probably holding other people up. Much to his pleasant surprise, a couple of his colleagues asked if they could help and then proceeded to carry him down four flights of stairs. Looking back at the incident, he

considered that the whole experience had improved the team spirit within his department because they had learned how to help and trust each other in a time of need. In some ways, it reminded him of the community spirit which had been engendered among the people who lived through the bombing raids during World War II.

- Another entry was from a female employee who said that her main concern was the possibility of the lights failing in an emergency situation. Apparently, she was terrified of the dark and always carried a small flashlight in her handbag, just in case. She explained how a traumatic experience in her childhood – being shut in a dark closet by accident – continued to trouble her. The employee said she was looking forward to the first evacuation drill because that would reassure her that it was possible to get out of the building safely in an emergency. In fact, she went on to say that she had actually practiced getting out of the building as quickly as possible, but was still worried that everybody else would get in her way if they all tried to leave at the same time.

A few weeks later, the personnel manager received a thank you note from someone who was pleased to have been moved to an office which was situated on the ground floor close to the main exit from the building. Clearly, the anonymous nyctophobe (nyctophobia = fear of the dark) had found a more comfortable position.

On the reverse side of each card was a map showing the exit points and the routes to the assembly areas. There was also a discreet notice in the waiting area which recommended that visitors should read the emergency evacuation instructions while waiting for their hosts to come and greet them.

## 2.4.1 Message Strategy

Once you have developed an outline strategy for your emergency evacuation plans and procedures, you have to start thinking about how to distill this all down into a simple message or series of messages. After that, you need to work out how to get those messages across using whatever means are at your disposal. I remember a time when the IT director of a small company had a series of cartoons produced to communicate the message that she was developing a disaster recovery plan, and these cartoons were pinned up on bulletin boards throughout the offices. Twenty years later I still have one of the originals hanging on my office wall acting as a reminder of the plans which I helped to develop and test. Every time I look at that drawing, it evokes clear memories of Yvonne Campbell and the work we did together. This one image links my mind automatically to a string of associated images, and it continues to do so unerringly after all this time.

Publicizing your EEP program is a relatively simple, straightforward marketing exercise. You know the identity of the target audience, you know that what you are trying to sell is something which will prove beneficial to every single one of them, and you know that the price is right. They are a captive audience, and all you have to work out is the means of communication and the style. So the question has to be: What vehicles for communication are already in place and how can you link into them?

## 2.4.2 Conveying the Message within the Organization

Common ways of communicating within an organization include the intranet; the in-house magazine, which might be a physical tangible document or it may simply be a delivery system; a bulletin board (or boards), which could also be purely electronic; e-mail; group meetings of various sorts; or straightforward word-of-mouth from manager to supervisor to operator. Lastly, there is the staff handbook which is usually issued by the human resources or personnel department and is updated from time to time.

Wherever there is a staff handbook, you should be aiming to have something distributed as an update or an addendum in the short term with a view to having the subject included in future editions of the handbook. The other route to a significant part of the audience of the future would be through the regular orientation process for new employees, which generally covers health and safety matters together with any related subjects such as ours. Again, it is the HR or personnel department that should be approached to deal with this type of activity.

If the company has bulletin boards used for internal publicity, then you should check that it is permitted for you to pin up a notice about EEP. While it is unlikely that anyone would object to the idea, it is polite and prudent to speak to the person who is in charge before taking action on what is, after all, that person's responsibility. It is also an opportunity to explain to him and her personally what you are trying to achieve. Perhaps an electronic bulletin board is available, in which case you can follow the established protocol to get your message displayed.

Wherever possible, I would like to enlist the services of a proper graphic artist to create an imaginative, attractive poster rather than simply rely on a plain text document. Of course if you or a co-worker happen to have a talent for this sort of thing, you can do it yourself without resorting to a professional artist. Whichever way you decide to approach it, the outcome should be a poster which portrays its subject in as attractive a manner as possible in order to capture the attention of the potential audience.

One of most memorable and meaningful posters for me has always been the one of Lord Kitchener during World War I, pointing directly at the onlooker saying, "Your country needs you." If I'd been there, I would have been tempted to join up straight away!

## 2.5 Aligning With Business Continuity

In Phase 1 of this book, we compared and contrasted BC and EEP and explored some advantages of communication between these two disciplines. That was only the beginning of your partnership with BCP. This process needs to continue throughout all phases of planning and exercising your EEP. Now that we are in Phase 2, we advise that if you find a successful BC program in place, then you should look towards the BC manager as a possible role model. On the other hand, if BC is not yet in place, perhaps you should see yourself as a pioneer. Because our lifecycle model is derived directly from the one used by BC practitioners and the outcomes are similar and complementary, it makes sense for you to consider aligning the two disciplines or even combining or merging them.

Possibly, you may find that a BC program has stalled or failed, in which case you need to find out what caused it to go wrong. It could be a clash of personalities, lack of funds, inadequate resources, a change of heart, poor management, or some other problem. Try to learn the lesson before embarking on your own voyage of discovery.

### Case Study: 10,000 Hours of Evacuation Practice

Sometimes, practice means everything. It's obvious in watching a soldier strip and reassemble a weapon, or professional athletes perform a move they've performed countless times. Malcolm Gladwell's, "10,000-Hour Rule" asserts that the key to success in any field is, to a large extent, a matter of practicing a specific task for a total of 10,000 hours. Why should successfully and safely evacuating a building be all that different, at least in person-hours?

Tippering Products, a manufacturing company in Ontario, Canada, took this recommendation to heart with amazing results.

Tippering operates a well-maintained primary manufacturing plant in a light industrial area. Because of the nature of the manufacturing process - involving welding, chemical baths, and acids - high safety standards are in place, and occasional area or building evacuations have been necessary for the 350 employees.

Procedures for orderly evacuation have been integrated into the corporate culture since the facility was built in the early 1960s. Evacuation drills are conducted several times a year - at group or department levels as often as monthly, and the entire plant, quarterly. Evacuations include emergency shutdowns and safety cutoffs of power, gas, and chemical flows via red "panic" buttons as employees exit. Workers are very conscientious about these details.

A comprehensive review of the evacuation procedures was conducted in in1999, in 2005, and in 2012, with another review in 2012. Most companies would find these long intervals to be a sign of neglect; on the contrary, Tippering's management team is proud that they can go a several years between formal reviews. Management has this kind of confidence because of how the emergency evacuation program is managed - not by managers, but by the shop workers and the people affected most directly by an emergency evacuation. Unlike those of many enterprises, Tippering's procedures are maintained by the people most vested in a safe evacuation.

A review of the Tippering change logs reveals an amazing level of maintenance activity for evacuation procedures. Virtually every section is reviewed and validated by more than one team many times a year. Data elements are updated and verified. External elements, such as routes, signage, contact information, etc., are all checked and cross-checked routinely. Besides scheduled exercises, smaller teams conduct their own walkthroughs and exercises. In short, the plan isn't simply up-to-date; it is more like up-to-the-minute.

Every last employee is familiar with the plan, and knows what to do when evacuation is necessary (whether or not a copy of the plan is at hand) because, at all times, they have been through an exercise very recently. Co-workers ensure that all new employees are walked through the evacuation plan before they are even trained in their job responsibilities, and newcomers are assigned an "evac" partner at least until they are fully certified as trained. Even visitors are given an evacuation mini-briefing and assigned an "evac" escort.

How does this relate to the "10,000 hours of practice" rule? If you calculate 30 hours per employee for about 350 employees, Tippering passed that 10,000-hour mark years ago. Having evacuation procedures part of the daily work life and company culture of employees has meant many successful and safe evacuations of the facility. Since some of these events very likely could have resulted in injury or even loss of life without training, the payback for that training is perfectly clear. What isn't clear is why any other enterprise would hesitate to make a similar investment in training for evacuation!

*Gladwell, M. (2008).Outliers: the story of success. Little, Brown and Company.*

## 2.5.1 What to Do When a BC Program Is in Place

If a successful BCM program or system is in place, you should seek to align your approach with the existing way of doing things. Of course, it is likely that they may not be using the same lifecycle model portrayed here, but you should still regard the overall principles, objectives, and outcomes from their process as a useful model which you could adopt as a prototype. Do not attempt to impose your way of doing things upon them unless you are actually invited to do so. If you are the new kid on the block, it is wise to begin by fitting in with what everybody else in the neighborhood says, does, and thinks.

**Eventually you might want to cooperate with their testing and exercising program, perhaps by including emergency evacuation as an integral part of some BC exercises.**

Obviously you won't be able to make your scheme fit in with theirs unless you have a pretty good understanding of what they have done, how they have done it, and where they think they are going. So, your first move should be fact-finding; ask them to tell you all about BC from their standpoint and how it has developed or progressed within the organization which you both share. It will be useful for you to understand how it all started and how it developed, since that might give you some clues about the best way to move forward in the same environment, dealing with the same culture and people.

An important outcome from this dialog will be an introduction to the tools and techniques which have been employed to build the business case, win support, gather the information, agree on the strategies, develop the tactics, and prepare the plans. You will also want to know about the ongoing awareness and training program which you can link into. Eventually you might want to cooperate with their testing and exercising program, perhaps by including emergency evacuation as an integral part of some BC exercises.

If it is not convenient or possible to merge the two types of practical training into one comprehensive and realistic exercise, then you should at least make sure that the staging of such events is synchronized so that people are not expected to take part in two similar activities at, or near, the same time.

The BC manager may want to include some of your ideas and plans within the set of plans for which he or she is responsible. This could be achieved in a number of ways. Your procedures might be included within the overall BC plan as a discrete section or module. They might also be added as an appendix or addendum to some or all of the plans. Another possibility is that the BC plans will reference your plans or even include a brief summary to make the reader or user aware of the EEP plans and procedures. If there is some merging of the two types of plan, then you must establish who is to be responsible for the review, maintenance, and updating of your sections, modules, or appendices. You also have to agree on a procedure to enable this to happen smoothly and in a timely fashion.

## 2.5.2 What to do when No BC Program is in Place

If you find yourself in the unenviable position of coping with an unsuccessful BC program, you need to find the best way to move forward. You have to choose between ignoring what went before, trying to prevent mistakes or avoid accidents, or attempting to salvage or repair the damage that has been done already. You can make such a decision only after you have established the facts, and this might not be too easy. Some people might want to avoid or withhold the truth, others may not know the truth, and possibly the truly guilty or innocent parties may no longer be around.

Try to find out the basics of what happened and why, but don't spend an inordinate amount of time and energy playing the role of detective. If the facts are to be useful and reliable, they should be relatively easy to uncover by asking around. But, if the truth is buried too deep it may never reveal itself. Once you have a clearer picture, or have chosen to let sleeping dogs lie, you can decide whether you want to embark on a rescue mission or plow your own fresh furrow. In either case, you need to set your priorities. Does the EEP program take precedence over the BCM program, or is it the other way around? Because human lives are at stake, I would recommend that you consider EEP as your number one priority, but be prepared for BCM to be raised as a subject by others. When, and if, the issue of the past BCM attempt is raised, you should explain your priorities to those bringing up the subject: ***We are pursuing the course of protecting the people before focusing on protecting the business.***

Perhaps you can refer to corporate policy, rules and regulations, or legislation in support of your argument and stance that people's lives are of primary importance. It would be a harsh person indeed who would wish to fault this line of argument; either that, or they simply don't, or won't, understand what you are trying to say.

## 2.5.3 What to Do If You Decide to Take on BC

Once you have completed your first cycle of the EEP program, you can return to addressing the issues surrounding the inactive BC program and, in the light of the experience and knowledge you have gained, you will be better placed to rekindle the BCM flame. Avoid trying to pin the blame for the past on anyone - now is not a good time to be making enemies. Simply accept the facts as they are and try to move forward towards a reasonable solution. By the way, you will find that there is never a good time for making enemies - they are an unnecessary burden which brings no benefits.

In the case where BC has not yet been established as one of the regular management disciplines, then you are most likely to be the candidate for leading the way in this desirable endeavor. I have no doubt that BC is a very worthwhile activity regardless of the organization or its circumstances. However, I do accept that you might consider it to be beyond what you are prepared to do, or might be allowed to do.

Assuming you are willing and able to launch this important program, you might want to pause and look ahead for a while. Do you see yourself taking on the role of BC champion, or would you rather see someone else running the program in the longer term? As we've already hinted, the experience of setting up and delivering an EEP program will set you in good stead for the similar but more complex tasks of running a BC program. Of course, the decision is yours and you must feel free to make up your own mind. Having decided to tackle BC as your next major career move, you will need to take stock, take a deep breath, and find out a bit more about the subject. Now is a good time to invest in some reading material or indulge in some training. Not only will the EEP experience prove useful as an example of the kind of work involved, but it will also have earned you the respect and admiration of those you help with that program. They will know that they can believe in you and will trust you, making the whole process go so much more smoothly.

## Phase 2 - Key Actions

- **Design the basics of an EEP program.** Ensure that it can be seen to be fit for its purpose, taking account of the environment and the local culture.
- **Conduct a training needs analysis.** Establish what needs to be taught and who needs training.
- **Develop the training content.** Anticipate long-range needs, since the training content will grow and change with each phase of EEP effort.
- **Establish when, where, and how you will deliver the training.** Liaise with those who plan and deliver existing company training.
- **Handle the logistics of the training.** Take into account the venue, scheduling, facilities, materials, presenter, catering, and protocol.
- **Plan and implement a publicity campaign for your EEP program.** Use all the means at your disposal to let everybody know about it. Anticipate long-range needs, since this publicity will become an ongoing part of company life.
- **Continue to align your efforts with BCM.**

## Discussion Questions – Phase 2

This is still an early phase within the EEP lifecycle, but it is an important one. This is the point at which you begin to make the EEP concept part of the "furniture," establishing the foundation which will turn what started out as a vague dream into a substantial reality. Your people should begin to understand and appreciate the benefits which can accrue from thinking ahead and planning for whatever the unknown and unpredictable future might hold.

1. In the text we have introduced the idea of using a memorable phrase to help people remember the essential elements of evacuation safety or emergency procedure. Can you come up with one for evacuation? Perhaps a catchphrase which matches, or relates to, your organization or its people?
2. What do you think is likely to make your people aware of the need for, and the benefits of, an EEP program? What is the message and how will you communicate it? Do you think the language is important, or is it simply a matter of bringing the self-evident truth to their attention?
3. How will you handle an apathetic or uncooperative response when you come across it? Will you invoke the power of the authority granted to you, will you try to argue the case from a logical point of view, will you appeal to their emotions, or will you refer to some aspect of the rules and regulations which apply?

# Phase 3

## Developing an Understanding of the Environment

At this third phase of the lifecycle you begin to gather the information that will form the basis of the planning, help you to understand the scale of the exercise, and provide you with factual input regarding the type and quantity of any required to support the evacuation procedures.

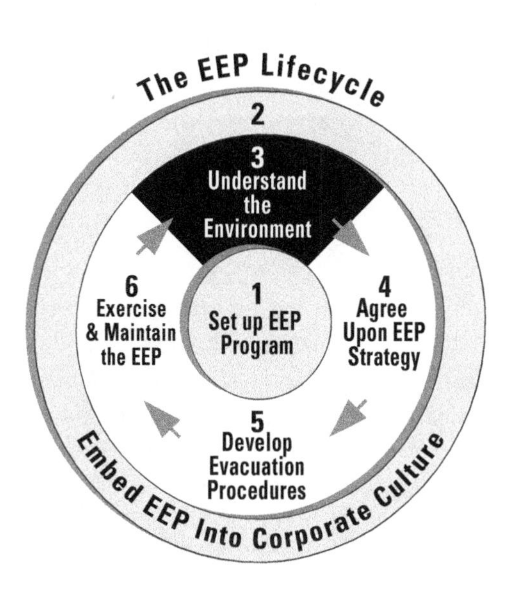


At this stage, you will need to do a good deal of research and cross-checking to get your facts right and have the evidence to support your proposals.

*This section will help you to*:

- Perform the data collection needed to determine the needs and requirements of persons at the site.
- Understand the terminology, concepts, and process of emergency impact analysis.
- Conduct physical and risk assessment of the site.
- Analyze the escape routes from the site.

Here at Phase 3 is the stage at which you will be doing the main exploratory substance of the work that you originally outlined and proposed in Phase 1. Before you criticize this as "thankless research," remember that you are performing this detailed fact-finding in the service of those people who are likely to be found in your environment – and for whose safety and welfare you will be developing a fool-proof EEP. It is this fact-finding phase which forms the basis upon which the EEP plans and procedures will be built. Remember the old adage from carpentry, "Measure twice, cut once." If you do the data collection right the first time, you will save yourself and your colleagues trouble and pain in the long run.

## 3.1 Data Collection Parameters

The method behind our EEP lifecycle relies upon a number of parameters, or concepts, which you will be using to guide you in your selection of strategies and tactics. We are now going to introduce new terms which describe the objectives, outcomes, and procedures of this phase of the overall process. These are all derived from the terminology associated with the original business continuity management (BCM) lifecycle model which inspired the development of this one.

MTPE = maximum tolerable period of exposure
ATO = assembly time objective
MTMP = maximum tolerable missing persons
AAR = assembly area requirements
SED = safe evacuation distance

These parameters can all be determined relatively easily through a process which I am calling an *emergency impact analysis* (EIA). You may also need to conduct an emergency resource analysis (ERA) in order to consolidate the special requirements which the EIA process may have identified. Considering the resources available would be particularly relevant where you are considering the needs of those with restricted abilities.

It is more important for you to understand and appreciate the process than to come to grips with the terms and their associated acronyms. It is the process which will produce the results and demonstrate that you are following good practice. Auditors, regulators, and inspectors will have a higher regard for you and your results if they can see that you have adopted a formal approach which takes account of all the possibilities and is fully inclusive.

- The *maximum tolerable period of exposure* (MTPE) is the absolute maximum time which can be allowed to elapse before everybody has reached safety. It is an estimate in the sense that it cannot be measured and it is a limitation in the sense that we do not wish to exceed it because that implies unacceptable levels of risk and danger. This MTPE should be set in collaboration with those who are responsible for the welfare of the people and activities in each location or occupied area.

- The *assembly time objective* (ATO) is the amount of time allowed for people to reach their assembly area safely. It is a target time which will ensure that you are able to evacuate everybody well within the maximum tolerable time period. Obviously, it should be considerably less than the MTPE; the closer it is to the MTPE the higher the risk. On the other hand, a shorter ATO may incur additional costs, require more complex solutions, or even be unattainable. These key parameters need to be set in consultation with those who have the knowledge and the responsibility regarding the people and their environment.
- The *maximum tolerable missing persons* (MTMP) parameter represents our target in relation to being able to account for everyone who was in the area at the time of the emergency. The ideal would be zero, in which case we could assure the emergency services that every single person was safe and accounted for. This goal is unrealistic and possibly even unachievable in many situations, but it is wise to find ways of accounting for as many as possible. The MTMP is the margin of error which we aim to allow ourselves.
- *Assembly area requirements* (AAR) is the term which describes the required characteristics of the assembly areas where we expect people to gather once they have left the premises. The definition will need to include dimensions such as distance from the exit point and capacity in terms of the number of people it can accommodate. You will need to specify other characteristics such as whether it should provide shelter, seating, or catering facilities. Accessibility also has to be determined, taking account of who owns the property and whether the owner controls access to that area. If space is at a premium in the neighborhood, then you may find that the emergency services will want to use the area for triage or as a base for their operations. These considerations would mostly apply in the case of a large-scale emergency, but then that is what you are supposed to be planning for – the worst case scenario.
- *Safe evacuation distance* (SED) is how far the people must be moved to ensure their safety and would normally be measured in terms of the distance from the point of exit.

## *Case Study: The High Cost of Good Intentions*

Valseca Corporation, an import-export business based in Mexico City, D. F., Mexico, maintains offices in a two-story building in an older section of the city. The original building structure is quite old, with wood framing. While the exterior is quaint and attractive, the interior is modern and corporate with an eclectic, Latin-American flair. The site is surrounded by the narrow and congested old streets of a previous era, which first led management of this 50-person company to develop basic evacuation procedures in 2004.

Because of abutting buildings on both sides of the building, the only usable exits are through the main doors in front and a rear door to a small parking area. A single staircase located toward the center of the building serves the second floor. The evacuation plan calls for employees on the second floor to use the windows on the east side of the building to exit onto the adjoining single-story building's roof if needed. Sad to say, the evacuation plan, that was created with the best intentions in 2004, was given only minimal maintenance after the initial evacuation initiative.

In late 2008, an electrical fire started around midday in some computer equipment which was right at the base of the central staircase. Unfortunately, this equipment was directly adjacent to a large, open file area, and the fire spread rapidly to some paper files. Billowing smoke spread, and flames surrounded the central stairway, making it impassable.

Employees on the first floor rushed to the two exits, pretty much ignoring any evacuation procedures. Since there had been little training, most were not really familiar with the documented procedures anyway.

Seeing the stairway blocked by flames, employees on the second floor began to panic. A few rushed to the front or the back windows; some actually jumped out onto a parked van outside a rear window. Several employees escaped out the side windows onto the adjoining rooftop, only to discover that in the years since that procedure had been written, that rooftop had been completely surrounded with fencing and barbed wire. In effect, they were trapped and were in far greater danger should the fire spread to that all-wood structure before rescue arrived.

A number of employees suffered injuries, although, fortunately, none were serious. Two employees subsequently refused to work on or ever go to the second floor again. One long-time, valued employee resigned. Litigation, productivity losses, and other direct costs were significant. Aside from property loss, business interruption, and other tangible losses, avoidable losses related specifically and directly to the failed evacuation amounted to over US $750,000.

## 3.1.1 Techniques

Once you have gained the support of top management in the form of a policy and have been given permission to proceed, you have to set up the rest of the program and begin to embed the ideas and principles into the organization's culture. Publishing the text of the policy will ensure that your authority is recognized. Here we are using the term publishing to imply distributing the text to everybody who works within, occupies, or visits the organization's premises. At the same time, you will also be continuing to develop the awareness effort, through which you will be embedding the concept of emergency evacuation planning into the organization's culture. (You already started to do this during Phase 2 of the lifecycle and will continue to educate everyone as the program develops and grows.)

In Phase 3 of the lifecycle, we begin to establish these important parameters which give a sense of scale to the whole process. To achieve this you will need to understand and use the relevant techniques, which include:

PRA = physical risk assessment
EIA = emergency impact analysis
ERA = escape requirements analysis

## 3.1.2 Physical Risk Assessment (PRA)

The first step is to conduct a PRA, which involves inspecting the premises with at least one other person who is familiar with the areas and the activities which take place there. Typically, this person would be a member of the facilities management team or perhaps the manager of that area. Obviously, if you get multiple viewpoints, you can have more confidence that nothing important has been overlooked. Remember that you are looking to identify the various threats which could trigger the need for an emergency evacuation. It is important to be thorough; lives could be at stake if you overlook or misinterpret something at this stage.

The outcome of this process will be a prioritized list of potential threats to the physical environment. They will need to be prioritized in the order of both their likelihood and the scale of their impact, which leads us to the next step, which is to conduct an EIA.

## 3.1.3 Emergency Impact Analysis (EIA)

For each of the risks identified in the PRA, you will need to develop an estimate of the potential impact on whoever will be on the premises at the time. Here you will need to work in collaboration with the local management team who will be familiar with the layout of the property and the people who are likely to be there. One must avoid getting bogged down in too much detail at this stage; so a relatively simple measurement scale should be used to categorize the impacts associated with each risk. *High, medium,* and *low* would provide an adequate degree of separation. High-impact would suggest that every effort should be made to implement a safe and speedy exit; medium-impact would suggest that such events need to be taken into account in your planning; while you might assume that low-impact events will be covered by the provisions you will be making in order to deal with the other two categories.

**The question of special requirements needs to be looked at from both a group perspective and an individual perspective.**

## 3.1.4 Escape Requirements Analysis (ERA)

Armed with information regarding the threats and their impact, you will need to establish what is required to deal with these events. I call this process an ERA. Working in concert with the same people and the likely occupants of the areas concerned, you will need to figure out their requirements for a successful evacuation. The factors to be taken into account will include:

- Likely volumes of people to be evacuated.
- Special requirements of those people.
- Safe evacuation distance(s).

Throughout this process, you should maintain a radically pesimistic rather than a conservatively hopeful view. If you plan for the worst case and a better case occurs, then nothing is lost; the other way around is likely to lead to disappointment or worse. Overestimate rather than underestimate the volume of people, but underestimate their likely rate of travel, making some allowance for unexpected difficulties or delays. Allowing for contingencies is another way of saying this.

The question of special requirements needs to be looked at from both a group perspective and an individual perspective. From a group perspective, you will be exploring the type of people involved: for example, a shopping mall would attract a wide variety of people of all shapes, sizes and ages; on a college campus, you might expect to find mostly teenagers; whereas a retirement home would probably house elderly and disabled people.

From an individual perspective, you need to find out whether the group includes, or is likely to include, those with special needs. If the answer is likely to be "Yes," then you need to explore the possibilities in regard to numbers and their likely whereabouts. Wherever there is a strong possibility that an evacuation will include people with special needs, you should arrange to develop personal emergency evacuation plans (PEEPs) with them as discussed later.

Once you understand the population to be handled in the event of an emergency, you can begin to determine what they will require in the way of assistance or protection during their evacuation, resulting in an initial ERA, which will need to be refined and improved during the development of the personalized plans.

You also have to determine how far all of these people must be moved to ensure their safety in the various circumstances highlighted by your PRA. This is the SED factor and has to take into account the likely scale of the trigger event and the extent of its consequences. For example, a collapsing building is likely to spread debris as far as the building is tall, i.e., a building which is 50 meters (or 164 feet) high, could cause a pile of rubble up to 50 meters (or 164 feet) away from its original base. In the event of an explosion, broken glass is likely to fly a considerable distance and is best avoided by hiding behind a barrier such as another building. This distance and the associated dangers are considerably reduced if the windows are covered with anti-blast film which prevents glass fragments from flying. The film binds the fragments together so that the sheet of shattered glass behaves more like a piece of heavy fabric than a shower of splinters. Thus the broken window tends to collapse in one piece under its own weight rather than take to the air in flying fragments. We would always recommend that all of the windows should be treated this way in any building which is likely to experience damage from a severe storm or a blast. Many suppliers and installers offer this type of material - a quick search on the internet for "anti-blast film" (or "bomb film" as it is sometimes called) should reveal a nearby source.

When you are estimating the SED, you should allow for the possibility of the emergency services imposing an exclusion zone around the scene of the trigger event. Ask your local police and fire service how, when, and where they would set up their cordons. The police in London, for example, have a general rule that the cordon should be set at about 500 meters (or 1,640 feet) from a suspect explosive device. This rule is derived from practical experience gained from the IRA bombing campaign during the 1980s. Of course, the size and shape of the exclusion zone would depend on the nature and scale of the incident.

The outcomes from the ERA process include an estimate of the number of people to be considered. These parameters might be called the Crowd Volume and the SED. You should also draw up a documented action plan to deal with the ongoing issue of producing personalized plans for all those who might need them. (In the example on the United Grand Lodge of England, we go into more detail about the dimensions of crowds and flow rates.)

Once you have carried out these investigations and collated your results, then you will have an understanding of the environment, enabling you to move on with confidence to the next phase, which is where you will use this information to help you to develop the EEP strategy (or strategies). It is important to make sure that all of your facts are in total alignment before exposing them to others' scrutiny or starting to use them as the basis for further developments. In other words, be careful and cautious with your collation. It is also important that you retain all this information and store it safely for future reference. You and others will want to refer to it when reviewing, auditing, or investigating the EEP program.

## 3.2 Physical Risk Assessment (PRA)

As we mentioned earlier, the first step towards understanding the environment and the needs of its occupants is to inspect the premises, looking for potential problems, in the company of someone who is familiar with the areas and the activities which take place there. This inspection is the basis of what we are calling a PRA. Throughout this stage you should keep your eyes and ears open for any information you can glean which may prove relevant or useful at a later stage. Be childlike in your approach and try to learn as much as you can by probing and questioning. You are on a gentle and broad learning curve at this stage of the game.

Your ideal companions and guides throughout this process will be persons who have been working in the area for a long time; this will mean they remember some of the problems from the past. They should also know most of the people whom you are likely to meet in their area; this will give them an insight into the particular needs and requirements of those individuals. Their knowledge of the environment such as the facilities, services, and equipment to be found in the area will be an important asset; they will understand the implications of faults and failures which might impact upon the people. Your escorts should also have access to all the nooks and crannies in the environment; this will enable you both to explore those areas and items which everybody else ignores, avoids, or is unaware.

Companions most likely to meet this description would be whoever is responsible for maintaining and servicing the resources and facilities in that area. These people are usually engaged in what is known as facilities management, although other titles may be used, such as estates management, service department, concierge, janitor, property manager, buildings and grounds, or caretaker.

If you are working in a large and complex environment, such as a multi-tenanted building, a business park, a university campus, an industrial estate, a shopping mall, or a retail park, you may need to elicit the services of a number of individuals in order to develop a clear and comprehensive picture of the environment and its implications. In such cases, a wise choice of companions will make life much easier and the outcome more reliable. Don't hesitate to ask for advice in your search for the ideal escort; marriages might be born in heaven, but successful partnerships usually spring from careful selection. The personnel or HR department might be a good place to start.

The risk assessment process is pretty straightforward in principle, providing you remain alert and pick up on all the signals which you and your companions come across during the exercise. Your first task is to explore each other's views and opinions as a form of introduction and orientation before you begin your assessment of physical risk. Discuss the way in which you will be working together, the tools and techniques you might want to use, and the outcomes which you are seeking. Once all of you know what is expected of one another, it will be much easier to cooperate and benefit from your combined strengths and knowledge. During this conversation you will be developing a mutual respect which will then form the basis of a sound working relationship.

**What is to be the scope, breadth, and depth of the work at this stage? It is a good idea to have a checklist or an agenda outlining what you think you will be looking for and where you will be looking.**

When I am working as a visiting consultant, it is during this first meeting that I begin to get a clear picture of the overall culture and ethos of my host organization. It sets the backdrop as we prepare to go on stage and act out our roles. By listening carefully and adopting an enquiring mind, I get a feel for whom and with what I am going to be dealing. Often we find common areas of interest which underpin our working relationship and allow us to engage in a deeper and more meaningful conversation. Bear in mind that you might need to explore all sorts of unknown subjects and issues as you move forward towards identifying the potential risks and hazards that surround you.

Before very long, you will need to establish which broad areas of investigation you want to indulge in. What is to be the scope, breadth, and depth of the work at this stage? It is a good idea to have a checklist or an agenda outlining what you think you will be looking for and where you will be looking. This agenda could be prepared in collaboration with your escort/guide or you may prepare one in advance of your initial get-together. Whichever way it happens, you have to have some kind of a game plan; otherwise, neither of you will know where to start and how to recognize when you are coming to the end. It doesn't have to be an in-depth, precise description of the work; it is only a set of guidelines but will, hopefully, point you and your companions in the right direction.

## 3.2.1 The Tour

Once you have a reasonably clear idea of what you are trying to do and what each member of the team brings to the party, then you can set out on your tour of inspection. This is where a clipboard and a checklist will come in particularly handy. As you are taken around the premises, you should be looking out for all the potential causes of any incident which might lead to the need for people to be prepared to leave the building at short notice and escape to a place of safety. If you have any doubt about the structure, its use, or its contents, then you should ask questions. On those occasions when your partner can't explain something or is unable to give you a satisfactory answer, then you should make a note and follow up the issue with someone who can help you. When you eventually find the answer, it would only be polite to pass the information on to your colleague; he or she will appreciate your feedback.

In the accompanying EEP Toolkit you will find a document which is a tool designed to assist you in the assessment of physical risks in connection with EEP. Add any questions which might be appropriate to ask within your working environment and culture. Apart from improving the effectiveness of the tool, the process of considering what should be taken into account and how to approach the enquiry is a serious part of your preparation to do an effective job. Remember that this book can provide you with only the broadest of outlines because your situation is unique and the various tools and techniques must be adapted to fit the individual site and its circumstances. Take plenty of notes: Memories can be incomplete or misleading and cannot be used as evidence or proof of having carried out the task. More importantly they form an audit trail which proves you have followed the proper process.

Once you have explored the place and asked lots of questions, you can start to assimilate the information gathered in order to produce some meaningful measurements or facts to feed into the next stage. What we are expecting to derive from this process is a prioritized list of physical risks which need to be taken into account and addressed. The results from the first tour may be inconclusive because the initial analysis appears to reveal inconsistencies, anomalies, mysteries, or conflicts. This means you will need to go around again asking more questions. Perhaps you will find it necessary, or at least advantageous, to speak to more or different people. The main thing is to be alert and thorough throughout this process.

## 3.2.2 The Outcome

Now that you understand the nature and scale of the risks, you need to develop a plan of action, outlining how those risks should be addressed or mitigated. Assuming that you have identified a number of them, they will need to be ranked in order of priority. If you elected to use the PRA Tool, which is in our EEP Toolkit, then you will already have been offered the opportunity to carry out this prioritization prior to preparing a report for submission to your sponsor. Do not be overly ambitious; select the four or five most important, and rank them for priority.

It is unlikely that top management will want to invest time, money, and energy in dealing with every little problem which is reported to them, at least not in the first instance. What they prefer are results which can immediately demonstrate their commitment to the important health and safety issues which you have raised. They might also be inclined to consider some of the more costly measures which you propose. The budget and resources for these solutions may not be immediately available, so you need to show patience rather than demonstrate frustration. Some of the more obscure or difficult measures may well be turned down, but do remember that this is an ongoing, iterative process and more progress might be made at a later date. You should end up with three or four types of recommendation in your report:

- Quick wins, for immediate, low-cost actions.
- Important issues, requiring investment, to be addressed as medium-term actions.
- Strategic issues, requiring major investment, to be addressed as long-term actions.
- Complex issues, which everyone should be aware of but for which solutions are unavailable, unaffordable, or impractical at this point in time.

It is not essential that you fill all of these categories, and it is even possible that you will need to invent or adopt another category to accommodate all of your ideas and suggestions. Wherever it is possible, give top management two or more options, making it easier for them to take a positive action. With two possible solutions, they can choose either of the options or do nothing, which gives you a two in three chance of a favorable response.

**Wherever possible, you should aim to err on the side of safety rather than cost or convenience – people's lives could be at stake.**

Your main concern should be to draw top management's attention to these issues, but they have to be allowed to take the responsibility for accepting and implementing your suggestions. At this stage, you are only the messenger; later, you will be taking on the role of planner and executor.

## 3.2.3 Risk Register

Having identified and assessed the various risks which you and others perceive, it is useful to log these risks to better track the progress of any proposed actions to deal with them. Such a log is commonly known as a risk register and it provides you with an overview of the current risk situation. Furthermore, it is a useful tool which you can use to demonstrate the situation or answer any questions which people might have in this connection.

For each risk, you should have a brief description of the risk and its rating; an indication of who is responsible for that area and what, if anything, they should be doing about it; together with a date for completion or review. A template named "EEP Risk Register" is included in the EEP Toolkit.

**EIA is an investigative technique with which we seek to establish the parameters or dimensions of the evacuation procedures designed to ensure the safety of our people in the wake of an incident.**

## 3.3 Emergency Impact Analysis (EIA)

The *emergency impact analysis* (EIA) process or technique is derived from the *business impact analysis* (BIA) process which is a key and integral part of a *business continuity management* (BCM) system. EIA is the stage of the planning process in which you set out to discover or develop the key parameters which are required to develop and select your EEP strategies and tactics. These parameters will also be reflected in the subsequent planning process.

Because the idea of an EEP lifecycle is relatively new, it has not been subjected to continuous evolution and improvement through experience and feedback. However, it is based closely upon the BC model which has matured through an extensive and rigorous development process. Therefore, we have great faith in the concept and the processes which support it. The only area where there may be some doubts or room for improvement will be in the terminology or the labeling of the various tools, procedures, and products involved. We have introduced meaningful and understandable words and phrases in this connection, though you may prefer to adjust some of the wording to suit your taste, culture, or working environment.

EIA is an investigative technique with which we seek to establish the parameters or dimensions of the evacuation procedures designed to ensure the safety of our people in the wake of an incident. They are not precise measurements which represent success or failure; rather, they are estimates or judgments which provide us with guidelines. Wherever possible, you should aim to err on the side of safety rather than cost or convenience – people's lives could be at stake.

An investigation can be conducted in many ways, and the final choice will depend to a large degree upon personal choice, the type of organization, the extent of the areas to be investigated, the number of people involved, the availability of facilities, and what is regarded as acceptable or common practice. You will need to take account of all of these factors when deciding how you want to proceed. I will describe some of the common ways of carrying out this type of research. You may come up with your own style. It is the output which is important, not the manner in which it was derived, although good recording throughout is advised to satisfy auditors, quality assessors, or any others who wish to inquire, review, or investigate at some later stage. These people will expect a paper trail which they can follow with ease.

The common ways of conducting this type of investigation within a business environment include the use of workshops, questionnaires, interviews, and various forms of inspection. You might find it best to combine two or more of these techniques to develop your own view and gather the information upon which to make a sound judgment. I often use a combination of all four methods before coming to any firm conclusions and submitting my recommendations.

## 3.3.1 Facilitated Emergency Impact Analysis

**Interactive Workshop:** For the impact analysis, an interactive workshop is probably the easiest way to gather some realistic, representative information and obtain a degree of consensus about what could or should be done to deal with the problems which have been discussed. I find it is the easiest way both to introduce the subject and at the same time to get a rough idea of where the various parameters might lie.

Invite a representative number of people from the location, the area, or the group which is to come under investigation, explaining what you hope to achieve. If there are a number of separate, or different, groups or locations to be analyzed, you can either adopt a hierarchical or a repetitive approach. With the hierarchical approach, you run a high-level analysis with top management to develop an overarching view of the problems and potential solutions before running further workshops with various groups to flesh out the details at a tactical or operational level. In a large or complex organization, you may even find it necessary to run some intermediate workshops with middle management before working out the fine details at the operational level with junior management or local supervisors.

While it is possible to run a joint workshop covering more than one situation, location, or group, you may find it difficult to keep the different perspectives sufficiently separate to prevent confusion. It is safer, and easier, to treat differing groups as separate entities which are worthy of your full attention.

At the start of the workshop, explain the context and what we are trying to determine. I try to make it absolutely clear to the participants precisely what it is that we are looking for and then we examine each of those parameters in the light of various credible emergency scenarios. Because they are familiar with the territory and its occupants, take their lead and allow them to suggest the possible emergency scenario with which they, their colleagues, and those they care for or support, may have to cope. If they should happen to miss out on a particularly relevant scenario then, obviously, raise it as another possibility for them to consider.

**Because they are familiar with the territory and its occupants, take their lead and allow them to suggest the possible emergency scenario...**

Usually fire is regarded as the most likely scenario they may encounter, since it is, by far, the most common cause of emergency in all types of property throughout the world. It is also the one which tends to spread its effects rapidly and therefore requires the best escape and evacuation procedures. Flooding is also a common threat in many locations, but it is rather site-specific and depends upon the local geography. Explosions and terrorist activity are also common suggestions for emergency scenarios, although in these cases we are usually using evacuation procedures as a reactive rescue strategy rather than a proactive protection measure.

In those instances where the structure of the building has been damaged by an explosion or similar event, then the emphasis has to be on cautious and careful movements to reduce the risk of injury rather than hurrying to reach a place of safety in order to avoid the risk of injury. This type of evacuation will be the subject of a totally different set of parameters because the primary consideration will have totally changed from reaching a distant safe space to that of avoidance of further danger and injury. It is also quite likely that damage to the structure and contents of the building may necessitate the use of different, or at least modified, routes and exits. Furthermore, the emergency services are likely to be in charge of what is now a rescue operation. The thought and work which has gone into the emergency evacuation planning process will be essential under these altered and difficult circumstances.

During these EIA workshops the two primary parameters which you are trying to pin down through consensus of opinion will be the MTPE and the ATO together with the concept of parameters in connection with EEP.

Another factor which should be easy to pin down at this stage is the likely number of people who will require accommodation at the assembly area. However, several dimensions and a number of other considerations have to be taken into account when describing or selecting an assembly area, which will be addressed when you conduct an ERA. For now, concentrate on deciding what is deemed to be tolerable rather than what will be required.

At some point you will confront the question of accounting for as many people as possible. The MTMP parameter will affect to some extent the degree of preparation and the style of management which you are going adopt in the event of an emergency. For example, if you decide to go for zero tolerance, it may be necessary to organize and train a large number of marshals and helpers to make sure that everyone is accounted for at all times and stages of the evacuation procedure. This may increase the number of people who are exposed to additional danger in the pursuit of safety for others. Another possibility is the use of some form of tracking device for every individual who is likely to be on the premises. However, although the use of a tracking device is technically possible it has found favor in few organizations.

There are no easy answers to this question of an acceptable number of missing persons, and so the issue is commonly avoided and assumptions are made, either knowingly or unknowingly.

Of course, it is to be hoped that everyone will be accounted for and your efforts will be concentrated on trying to ensure that this is the case.

## 3.3.2 Other Tactics

**Questionnaire:** Another popular form of inquiry used to establish, or help to establish, your core parameters would be a questionnaire. Given the difficulty in gathering accurate information, make sure that everyone takes the matter seriously and take pains to ensure that everyone fully understands the purpose and importance of the survey and questions. One way of enhancing the understanding is to run a short preliminary workshop in which you explain everything to them in advance of the actual survey.

**Interview:** Our final investigation technique is to conduct an interview. Sometimes this can be done over the phone or even by Internet, although the most effective interviews are those in which the two parties meet face to face. This is certainly the best way to get reliable information, but it is also the most time consuming. However, if the information is important, then it is worthwhile investing time and effort to obtain it. Usually such interviews are based on the use of a questionnaire or a checklist, and it helps if the interviewee has an opportunity to review the questions well before the interview. Try to meet at least three people to obtain a rounded view of the problem and the possible solutions. If you interview only one person, you might get a biased or limited view. When I speak to a second person from the same area about the same problem, I might get a very different view. Whichever approach you take for your information gathering, you will be increasing the awareness and publicizing the EEP program by involving all these people in the debate about the size, shape, and nature of the plans, procedures, and resources which you will be developing or providing on their behalf.

### 3.3.3 An EIA Checklist

A questionnaire from the EEP Toolkit can be adapted to suit your needs and your approach to this stage of the program.

The checklist is designed to allow for up to three persons to be interviewed in connection with each area of activity under investigation, and you can extend your inquiry out to a larger number if required. For each parameter, gather a number of viewpoints and then obtain a consensus based upon the various suggestions received.

In addition, the questionnaire offers you the opportunity to gather information which is the subject of an ERA. These procedures are often rolled into one in order to save time and effort because they usually require input from the same, or similar, group of people.

**Here we are trying to determine the requirements of those people in terms of a safe separation distance and the space to accommodate them.**

## 3.4 Escape Requirements Analysis

This *escape requirements analysis* (ERA) exercise is usually performed concurrently with the EIA procedure as a matter of practical convenience. We are describing them separately to make it easier for you to get a clear distinction between the two activities. Whereas in the previous phase we spoke about the gathering of information regarding the need to get a number of people safely out of the building, here we are trying to determine the requirements of those people in terms of a safe separation distance and the space to accommodate them.

While the questions you need to ask at this stage are different, the means of inquiry can be the same. Indeed, we would normally expect the two activities to be carried out concurrently, unless there is a good reason to separate them. In a multiple occupancy building, for example, the question of where to go and how to get there may be dealt with separately for each occupying group, although determining the overall dimensions of getting people safely out of the building may be seen as a joint exercise.

There may be other situations in which the two processes can, or should, be dealt with separately because of differing needs of various sections of the target community. Only you can be the judge, based on your knowledge of the culture, the environment, and the population concerned. Of course, it might also be possible that diplomacy or politics will be factors which you need to take into account when working out the best way to tackle this phase of the EEP program.

The two main characteristics which you are trying to establish at this juncture are:

1. The separation distance from the location of the incident to a place of relative safety
2. The size of the space required to accommodate the evacuees safely.

These two parameters can be determined as secondary or supplementary questions during the EIA. However, you will probably require much more detail to enable you to select the best and safest escape routes and assembly areas.

**Safe Evacuation Distance:** When estimating the safe evacuation distance (SED), you will need to make certain assumptions including the scenarios which might trigger the need for evacuation. Some events such as major flooding or pollution may require long-distance evacuation, which is probably beyond the scope of our current remit or area of activity because the response to this type of incident would normally be announced, managed, and controlled by the emergency services. Very often this type of event can be predicted, as in the case of a hurricane, for example, and there is time to organize and conduct a large-scale evacuation in advance of the actual incident.

What we are concerned about is the more localized incident which affects our premises and our people such as a fire, flood, or explosion. Here we need to consider such factors as flying debris, secondary fires, and shelter from the elements. There is also the question of localized, specific hazards due to other activities which take place in the neighborhood. Examples of such activities may include mining or other hazardous operations; dangerous or flammable materials which are held in storage nearby; chemicals in transport; or industries which may cause pollution or contaminate the environment. Local knowledge is essential in this area. Do you know what your neighbors do and whether there are any hazards associated with that type of activity? It is well worth inquiring. Your local authority should have a risk register which lists any such hazards; speak to the local emergency planning officer.

**Default evacuation distance:** As we have suggested elsewhere, the default evacuation distance is often determined by the emergency services, whose workers are likely to place a cordon around the area in the event of any incident which requires evacuation of all or part of the population. Typically, this exclusion zone is likely to be anything up to 500 yards, or 457 meters, from the scene of the incident. It seems reasonable to use this distance as the basis of our search for an assembly area.

**It won't take very long to draw a circle on a map to locate the likely venues which are worth exploring as possible evacuation areas.**

It won't take very long to draw a circle on a map to locate the likely venues which are worth exploring as possible evacuation areas. I find Google Earth is a useful tool for identifying open spaces which appear to be accessible. Obviously, one has to visit and physically inspect the place to check whether it is actually suitable, large enough, and accessible to the public. When carrying out this type of check, you do need to make sure that other people have not already designated this area as their emergency assembly area. If they have, there could be a conflict unless there is enough space to accommodate all concerned. The best way to find out is go and knock on the door and inquire – you might even make friends in the process.

**Needs of the people involved:** Consider the range of people who may need to be evacuated, taking into account their general and special needs regarding the escape routes which may be available and suitable for their use. For example, people with wheelchairs may not be able to negotiate stairs or steep inclines, and those who are frail or disabled may find it hard to cover difficult terrain without assistance. Our emergency evacuation checklist in the EEP Toolkit should prove useful in helping you to make judgments about these matters.

Local knowledge and an appreciation of the needs of the people involved will enable you to develop a checklist of the ideal or essential characteristics of suitable assembly areas and will also help you in the choice of the associated escape routes.

The kinds of things which need to be taken into consideration are those which affect or influence the welfare, safety, and comfort of the various classes of evacuees. It is worth bearing in mind that the evacuees may be expected to remain in that area for a considerable length of time, at almost any time of the day, at any time of the year, and in any kind of weather. Therefore, the list of ideal characteristics might include:

- Access to toilet or washroom facilities.
- Shelter from the rain, wind, or sunshine.
- Heating or cooling, depending on the weather and climate.
- Seating, especially for the elderly or disabled.
- Drinking water.
- Refreshments.
- Information about what is happening.
- Communication facilities, i.e., how are you going to keep them informed about what is happening and what is expected of them.

You might also consider the need for some form of entertainment or amusement while the evacuees are waiting and access to shops and other services, such as cash points or automatic teller machines (ATMs).

The result of your inquiries in this area will be a kind of shopping list to help you select the most suitable places where your people can gather in the event of an emergency which requires evacuation of the premises.

## Phase 3 – Key Actions

- **Review and understand** all the new terms used in this data collection phase.
- **Collect data.** Make sure you determine the needs and requirements of all those who may be on site at any time.
- **Conduct a Physical Risk Assessment**. Identify potential problems, hazards, and safe areas.
- **Compile a Risk Register** as a formal record of the physical risks.
- **Perform an Emergency Impact Analysis.** Determine all the implications of these risks.
- **Prepare a report.** Outline your findings regarding the impact of any risks which appear in the Risk Register.
- **Conduct an Escape Requirements Analysis.** Determine the needs and requirements of the people in regard to evacuation and safety; establish the volumes, and the minimum safe distance from the property.

## Discussion Questions – Phase 3

Because of the significance and the relative uniqueness of the procedures and processes involved in Phase 3 fact-finding, as you determine the needs and requirements of the people for whom the EEP is designed, it is a good idea for you to spend some time reflecting on whether you fully understand and appreciate what you are getting into. Perhaps it would be useful to try to explain the process, and how it is performed, to someone else. This will give you the opportunity to put the ideas into your own words and check that it all makes sense.

1. Are you familiar with any, or all, of the laws, guidelines, and standards which might be issued by local authorities, regulators, networking groups, trade bodies, or trade unions? Could you, through your colleagues or your organization, have access to any groups or societies which provide occasional advice and guidance to their members?
2. Are there any guidelines available to you that may be relevant to your particular type of industry or location? Are you a member of a professional body which might be a source of useful information in this connection?
3. Can you already begin to envisage the manner in which you will approach this body of work? Do you have some idea of the time frame involved? Perhaps you can try to calculate the work-hours involved and estimate how and when you will be able to gain the cooperation of your colleagues? What milestones would help you monitor your progress with this work.
4. You cannot, from a practical standpoint, explore the needs the Phase 3 requirements for EEP entirely on your own. Who can you identify as good candidates for your "team"? From what departments should they come? What backgrounds and skills would be useful? What should be their experience with the facility? How will you explain your research needs to them (or to their supervisors if necessary)?
5. Re-read the case study about the Mexico City factory. What lessons have you learned from this section of the book that could prevent a failed evacuation of this type from ever happening in an organization for which you work?

# Phase 4

## Determining Evacuation Strategy

By now in the development of your EEP program, you will have gained support for such a course of action, begun to embed emergency evacuation thinking into the culture, and developed an understanding of the environment in which the evacuation procedures will need to operate. Within our lifecycle model, you have reached the beginning of Phase 4, which calls upon you to start building upon the information you have gathered so far.

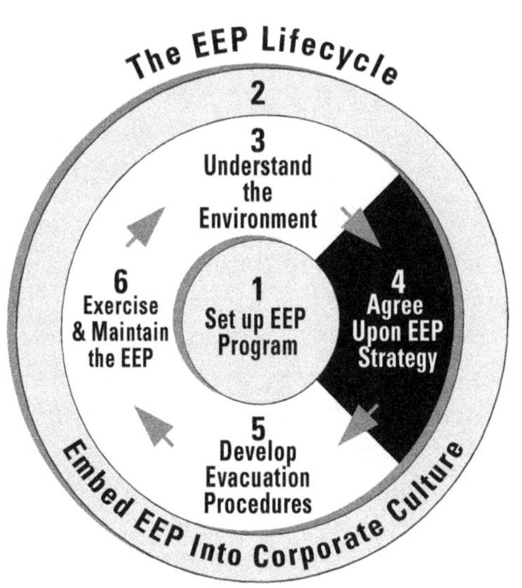


*This section will help you to:*

- ➢ Analyze environmental and population data, and review escape routes and assembly areas.
- ➢ Create plans and escape techniques for the disabled.
- ➢ Understand the EEP challenges of high-rise buildings.
- ➢ Create effective signage to support your EEP.

## 4.1 Gathering a Dimension of Environmental Data and Analyzing Necessary Information

### *Three Aspects of Information Gathering*

Three distinct aspects, or dimensions, of information gathering are involved when you are conducting your research in connection with emergency evacuation to gain a full and rounded grasp of the subject:

- Environmental data.
- Population knowledge.
- Education and learning.

**Dimension of environmental data:** The principal area of interest, which was the main focus of Phase 3, is to gain an understanding of the physical environment in which you are working. You sought to gain a thorough understanding of your surroundings in relation to the safety and movement of people. This dimension of *environmental data* – or knowing where you are – is all about looking and exploring.

You will already have gathered a great deal of the basic environmental data during Phase 3 as you carried out the physical risk assessment and the emergency impact analysis. We will amplify that as we survey the potential escape routes and assembly areas in order to develop a comprehensive understanding of the full environment.

**Dimension of population knowledge:** To make the best use of this environmental knowledge, it is essential to appreciate the needs and expectations of your target population, that is, the population which resides in, works in, or visits the premises in question. What's needed might be called "insider knowledge." In all probability, you already know who these people are. It might even be that it was their vulnerability which triggered the thoughts of EEP in the first place. However, a number of implications and aspects may not be immediately obvious and reveal themselves only when you begin to discuss these matters with people at the site, as well as the people who help or support them. A population survey involves finding out whom you are dealing with and what might be required to help them reach a place of safety. This survey information is important because you do need detailed knowledge of their needs, limitations, and requirements in order to figure out how, when, and where they will have to go in the event of the sort of emergency which requires them to seek shelter or move away.

**Dimension of education and learning:** Once you have gathered this information, you get to become a subject matter expert, someone who has a basic working knowledge of such matters as techniques, regulations, and methods. You need this sort of information in order to ensure that what you propose and promote is aligned with local, national, and international legislation; standards; and good practice. This information could be called the dimension of "professional data," which is all about knowing what other people say, do, and recommend in this connection. To gain this knowledge, you have to do some reading and remembering.

## 4.1.1 Site Review Process

A site review for emergency evacuation planning includes the whole of the neighborhood and the neighbors. You will need to identify any risks to people's safety throughout their progress from their place of work to the probable points of safety. Bear in mind the likely circumstances that would trigger the emergency evacuation and the number of people involved. Many of the potential threats would apply to most, if not all, of the buildings in the immediate vicinity.

**Exit Points:** Pay particular attention to the exit points from the building. Locate and identify at least two emergency exits, each offering different aspects or escape routes. Ideally, people should be able to exit in any direction, i.e., through a front, rear, or side exit. Where appropriate, each exit point should be equipped with a stout canopy or covered walkway to protect people from falling debris as they move away from the building.

Have a checklist ready when you are carrying out such a survey, especially if this is your first attempt, since it is easy to overlook some of the clues. There can be good clues and bad clues in the sense that some of them might reveal good news while others may suggest areas for improvement or even places to avoid.

**Once you have identified some safe spaces, you need to plot some safe escape routes from the exit points to the external assembly areas.**

**Safe Spaces:** The second object of the site review is to identify a number of safe spaces to be considered as potential emergency assembly areas. Your goal is to locate two types of assembly areas: *internal refuges* and *safe open spaces.*

- An *internal refuge* must be within the core of the building and not exposed to any external windows, since flying glass is one of the key dangers in any emergency. Internal refuges must also be proof against internal damage to the building. Seek the advice of a structural engineer in this connection.

- To protect against flying debris, *a safe open space* will be at some distance from the home location, not be in line of sight of the likely target, and a safe distance from nearby buildings. In practice, the area should be at least 500 yards away or within about five minutes walking distance.

Once you have identified some safe spaces, plot some safe escape routes from the exit points to the external assembly areas. Identify at least two alternative routes to each assembly area in order to avoid unexpected crowds, obstacles, or additional dangers. Ideal escape routes avoid straight lines because corners provide protection.

**Safer Indoors or Outdoors?** "To flee or not to flee: that is the question: whether 'tis nobler to stay behind and suffer the slings and arrows of fortune or to take a chance amidst a sea of rubble?" Hamlet clearly had a dramatic moment and was unable to make up his mind. In an emergency situation you have no time for introspection or speeches – you must decide and act accordingly without delay.

Establish whether your building can be considered as a safe refuge in the event of an emergency. If there is enough really safe space within the building, then invacuation is one of the options with the advantage that people need not expose themselves to any external dangers as they escape. This option has to be balanced against the possibility of being trapped inside the building. Clearly, if there is sufficient notice of the impending danger, then complete evacuation to a remote assembly area is the preferred strategy.

If the warning period is uncertain or very brief, then risks are associated with an evacuation. The event may occur while some or all of the staff are still en route. Making the best decision at the earliest moment is a key consideration in developing your decision process.

### 4.1.2 Assembly Area Requirements

To identify the correct assembly areas, begin by establishing the needs, requirements, and expectations of the evacuation crowd at the individual and group level.

Before you start the search for suitable assembly areas, establish the required characteristics so that you are able to recognize an appropriate area when you see it. Here we shall be describing the ideal assembly area, although you may have to compromise. Unfortunately, we all have to live and work within the environment as it exists rather than the ideal over-optimistic world in which everything is always available and nothing ever goes wrong.

Find a location that will safely and satisfactorily accommodate your population for hours at a time and at short notice.

The first thing you must establish is the amount of space that will be required, which is determined by the size and type of that population. It should be relatively easy to estimate the total number of people who are likely to be present when an evacuation is deemed necessary. You should take into account the possibility of visitors and passers-by swelling the numbers in the crowd; so err on the side of caution when it comes to determining the size of your potential assembly areas.

In the emergency evacuation checklist we suggest that you should aim for a ratio of one person per square meter (or 10.7 square feet) wherever people are expected to be able to move around in relative comfort. If this is unachievable, we suggest the population density might be safely doubled which means up to two persons per square meter (or 10.7 square feet).

People with special needs may require more space than this. For example, people who are in wheel chairs will require an absolute minimum of 1 square meter (or 10.7 square feet) each, which is actually quite cramped. Preferably, you should allow 1.5 to 2 square meters (at least 20 square feet) for each wheelchair-bound person in order to allow the person to maneuver easily and safely. The best way to establish the space and other requirements of those with special needs is to talk to them or whoever is responsible for looking after them.

Once you know the size and constituency of the evacuating crowd, calculate the amount of space required to form an assembly area. If you find no spaces large enough to accommodate the full crowd, consider multiple simultaneous assembly areas, which could lead to the additional problem of organizing and managing the various flows of people to prevent confusion and accidental overcrowding.

Apart from a space of the right dimensions to hold the right number of people, an assembly area must have a number of other characteristics, which will become apparent when you survey the potential sites and the associated escape routes. Looking at these other characteristics is the next step.

**Escape Route Survey:** During this stage of the EEP program, you will be exploring the various ways and means by which people can reach safety in the event of an emergency situation. You need to explore two interrelated areas in your survey:

- Where are people going to go in an emergency?
- How do they get there?

In the previous phase, our ERA identified the possible places for use as assembly areas. At the beginning of this phase, you assessed the assembly area requirements in terms of capacity and characteristics. Now you must investigate and compare those sites so that you can make a recommendation to your sponsor with regard to selecting, or negotiating for, the assembly areas of choice.

Ideally, you will identify a broad selection of such places, but it all depends upon the neighborhood and its occupants. At a minimum, to ensure safe access no matter the cause and location of the trigger event, you will need at least two options, in different directions from the target area.

**Your first challenge is to identify several possible assembly areas where people might be able to gather in an emergency.**

To make a selection, you have to engage with those who are responsible for, and familiar with, the territory in and around the premises with which you are concerned, including security people, facilities and maintenance, owners of the property, and health and safety personnel.

Using the emergency evacuation checklist from the EEP Toolkit, your first challenge is to identify several possible assembly areas where your people might be able to gather in an emergency. Ideally, I would like to have a number of alternatives for each point of the compass, so that we can be prepared to deal with a threat from any direction. Of course, this is the ideal, and a certain amount of compromise is to be expected, but you should identify at least a couple of alternatives.

Once you have figured out the possibilities, check them out thoroughly to determine whether they are realistic options. This reality check means talking to the owners of the property and establishing whether your people would be allowed to congregate there in an emergency and whether the area is earmarked for any other purpose in an emergency. It is quite possible that other organizations may share your ambition to use a particular area as a safe space.

You also need to consider how you can ensure freedom of access at very short notice and at any time. Get a written agreement about the terms and conditions of use, especially if your assembly area is somewhere other than public property. Your company secretary or legal adviser should be consulted before you finalize any agreements or contracts in this connection. Even with public property, it is worth checking that nobody else has designs on using that space for some other, or similar, purpose. Over time, you will also need to make regular checks to ensure that the space does not change hands, in which case you might need to renegotiate your rights to emergency access. It is also possible that the site may be developed or be subject to a change of use which could impact your plans in relation to declaring that property to be a safe haven for your people.

You may be able to identify an opportunity here to work in collaboration with your neighbors and enter into a mutual or reciprocal agreement whereby you can work together to your common advantage. Such cooperation could include taking part in each other's exercises or even embarking on some joint exercises. The ultimate development in this connection would be a group neighborhood safety scheme in which a number of organizations work together to ensure the safety of their people in the event of an emergency. I have seen this work effectively in a business park, a trading estate, and several retail complexes where it was seen to be to everybody's advantage to do as much as possible to ensure the safety and welfare of the public.

Having identified your assembly areas, you should explore the associated emergency exit routes, emergency exits, and the proposed escape paths, checking them out with reference to the emergency evacuation checklist. Make a careful note of any potential snags and any areas where additional aids such as handrails, barriers, signals, or signage might be required.

The outcomes from this stage will be a *final list of requirements*, which will include estimated costs, designated assembly areas, and clearly defined emergency exit routes, emergency exits, and escape paths. This list should be accompanied by an *action plan* or a schedule of works outlining how you will finalize the procedures and prepare the way for your people to evacuate in the sound knowledge that they will be able to reach a place of safety without any unexpected obstacles delaying or preventing their progress. All of this activity should be carried out in line with the *emergency evacuation policy*, making it relatively easy to obtain sign off approval from the program sponsor.

From a lifecycle perspective, the understanding of the environment has been completed, and now you are in the process of finalizing the strategy.

## Case Study: One Size Does Not Fit All

HALR Inc., a multinational financial group, operates from headquarters located on 18 floors of a 35-story building in London's financial district. Many of the company's mission-critical business operations and key employees are based there, although HALR has almost 20 other offices in 14 countries as well as smaller, satellite branch offices.

HALR's business continuity program (BCP) was built around a software package brought in by a contracted consulting firm in early 2005. Emergency evacuation was never an explicitly stated objective of the consulting engagement. Since the software package included a rudimentary emergency evacuation plan, the consultants expanded upon it somewhat by filling in the blanks and customizing a bit to the actual headquarters facility and organization characteristics. Although the BCP was exercised, the evacuation plan was exercised only minimally – in fact, the facilities department continued to run the same standard fire drills as always, unaware that an evacuation plan had ever been drafted.

Beginning in 2007, the BCP was rolled out to the 20 or so other offices, and scaled-down versions to the smaller, satellite offices. The same consulting firm was tasked with adapting the BCP to each office. Of course, this was a substantial project, but top management commitment to business continuity was strong, in part owing to increasing regulation of the financial industry worldwide. By mid-2008, the plans had been rolled out to every office.

On the other hand, emergency evacuation was, from the beginning, an afterthought that had been dragged along mostly because it happened to be included as part of the software, and not as an explicit mandate. Moreover, these offices ranged from three to eight floors of mid-city office towers to single-story, suburban-office-parks to strip-mall storefronts, with staffing ranging from 150 to as few as ten employees. The boilerplate evacuation procedures from headquarters (the plan that had been adapted for the London financial district) were pretty much dropped into each office, and employees were told these procedures were to be followed.

One of the smaller branch offices is located in a suburban location in Germany. A single-story, standalone building housing around 40 employees, it is relatively isolated in a quiet, wooded area. Local management of this branch, like management of other HALR branches, takes BCP very seriously. Consequently, when the consultants tailored the BCP program to this branch, local management reviewed it carefully and ensured employees were trained and conducted exercises. To some degree, local management tried to tweak the evacuation procedures to make them workable, but without any guidance, training, or resources, they assumed that headquarters and the consultants they sent were the experts, and relied on their guidance.

For better or worse, the exercises occasionally included the rudimentary evacuation plan. The consultant had not conducted any serious assessment to address evacuation; so the evacuation plan was pretty much boilerplate, and that boilerplate was pretty much appropriate to a 35-story building in London's financial district, not to a single-story building in the dense woods of Germany.

In early 2010, the evacuation plan was put to its first real test. Early spring weather led to excessive snowmelt, and a nearby river overflowed its banks for the first time in many decades. A steady, drenching rain over several days contributed to saturated grounds and drainage. Water levels rose within hours to the building's parking lot, within yards of the building's foundation. The business continuity plan was activated, and management was confident in the ability of the office to deal with the disruption; safety or evacuation didn't seem to be significant concerns, since the water level was at least seven or eight feet below a point where an evacuation would be needed, although employee cars were relocated as a precaution.

Late morning, suddenly all this changed. An extremely loud booming noise was followed by sparking, black smoke and flames from the side of the building facing the river. The main electrical transformers serving the building, which were underground, were fed by cables through a service tunnel which apparently had flooded. At this point, the light rain had turned into a deluge. Storm drains were overflowing, including a large catch basin which was close to the transformer – and not far from the rear emergency exit from the building.

While much of the evacuation procedure was boilerplate and pretty much useless, the basics were sound – particularly getting employees out of the building. Unfortunately, around 12 of the employees were directed to exit from the rear emergency door. As the first few stepped outside, the rushing water unexpectedly knocked them off their feet. One nearly drowned, but at the last moment, a second employee grabbed him to safety. Three employees suffered moderate injuries.

Once all of the employees were out of the building, the next step was to proceed to one of two assembly points of which, in this case would have been either under water or very close to that transformer. No other alternate location had ever been designated. Several employees figured they would be safer in the woods, which was clearly not a good idea in a thunderstorm – fortunately, there were no lightning strikes nearby.

After almost an hour to gather all the employees, there were 40 or so frightened, drenched, miserable employees, standing in the rain, some with minor injuries, unable to get to their cars or back in the building, with no other shelter. Other than calling emergency services, nobody had any idea what to do, with the exception of one employee who had a notebook computer with a copy of the

BCP, including the sample evacuation plan provided with the original software. With the help of a couple of others (to keep the computer dry!) they read and reread the evacuation plan – which clearly and explicitly directed them to *"proceed to the designated assembly location and if the primary assembly location is unavailable, proceed to the alternate [insert location here]."*

Further, it instructed, *"In the event this is not practical, find a local hotel, coffee shop or other establishment where you may all remain safe, comfortable and secure, and utilize commonly available communications and internet services, until building access is restored. The nearest facilities are [insert locations here]."*

*Postscript:* One year later, several of the trees in the forest surrounding the office building still bore signs labeling them "coffee shop," "Covent Garden Hotel," and so forth. One, in particular, to this day bears a sign reading "[insert location here]."

### 4.1.3 Assembly Area Assessment and Selection

The only way you can be sure that some of the prospective places are viable emergency assembly areas is to visit them and contemplate exactly how they will be used by your crowd of evacuees. Your first check will be to ensure that the dimensions are correct. In other words: Is it big enough for us all to gather in safety and comfort? You are looking for a number of features or possibilities which have to be taken into account at some stage. If you expect crowds of people in excess of a dozen or so, then you will need to establish a marshaling point where people are supposed to report and receive instructions, advice, or directions. Preferably, there will be a feature which can be used as a kind of podium that will permit the marshal to stand above the crowd so that he or she can be easily identified. Once you have satisfied yourself that the dimensions are adequate, get in touch with the owners of the property or their agents to answer your questions.

**Take into account any possible alternate uses of this property, occasional or regular.**

Of course, it has to be a safe space; here we might distinguish between a *safe open space* and a *safe sheltered space*. While a sheltered space is to be preferred, it could restrict the choice and also implies questions regarding accessibility at short notice. The ideal is unrestricted access in the sense that there are no restrictions affecting your use of the place at anytime. This is the point at which discussions with the owners or their agents should prove fruitful. Seek formal permission for emergency access, which requires that you understand the likely terms and conditions. Where terms and conditions are involved, ensure that the right level of representation is involved from both parties when thrashing out the details.

If there is competition or contention regarding emergency access, then you may have to refer the problem to a higher authority for resolution. Here I am using the term *higher authority* to refer to your top management, who might be better placed to negotiate on behalf of your organization. Of course, if there are other places available nearby, you should be able to reach a compromise by the simple expedient of eliminating the contentious space from your list of options. Don't forget to advise the other party of your decision to withdraw from the competition; otherwise, they may extend their search and cast their eyes on your alternate choice.

It is possible that a charge may be imposed for the use of someone else's land; in that case, some kind of formal agreement needs to be drawn up, outlining the conditions of use and the terms of payment. The owner may want to charge a fixed fee as a kind of service charge, or an hourly or daily rate could be imposed in relation to usage. It is also possible that there may be a combination of a long-term annual or monthly charge for right of access together with a usage or entrance fee of some sort. Obviously, you will need to ensure that these issues are resolved at the right level on behalf of both parties. This argument applies to all the questions in relation to the selection and implementation of any aspect of your assembly area and the associated escape routes.

If the assembly area is a currently undeveloped site, there is always the possibility that someone could set about developing the site. With a developed site, they could decide to change its primary use, which might render it unsuitable or unavailable for your purposes. You have to plan to review suitability and availability of all your designated emergency assembly areas on a regular basis. It could be most unfortunate if you discovered the answer to this question in a real emergency.

An important characteristic of an assembly area is freedom from traffic. It cannot be considered to be a safe space if traffic is allowed to pass through, and this applies to all forms of transport. Traffic to be avoided includes rail, trams, buses, lorries or trucks, motorbikes, bicycles, and even pedestrians. Try to avoid areas which include regular walkways or paths or at least make sure that your people do not obstruct the routes of pedestrians. Obstruction could lead to conflict, and in the worst-case scenario, someone could get hurt, physically or emotionally.

The final consideration in relation to the assessment, selection, and use of an assembly area is whether you can obtain permission to erect necessary signage on the site for the benefit of your evacuees. Such signage would include indications of meeting or marshaling points together with any direction signs which might be necessary or considered helpful. You might also need warning signs pointing out areas which should be avoided or left clear. Ideally, you should be able to erect permanent signs, but you may be able to get permission for temporary signs only, in which case your plans must allow for someone to take responsibility for managing the storage of these signs, as well as their quick delivery and display.

### 4.1.4 Characteristics of the Ideal Escape Route

Much of what we have said about the assembly area will apply to a greater or lesser degree to the assessment and selection of *escape routes*. Because an escape route is defined as the means of access from an exit point to an assembly area, both the start point and the end point are more or less predetermined. Your task at this stage is to figure out the best pathway between these two points and plot it. To achieve this result, conduct a careful survey of the various possible routes and determine which candidates are most suited to your purpose.

During the investigation, you should also be taking careful note of any improvements or modifications which would improve the suitability, or ease of use, of the various routes under consideration. Determining what modifications might be needed means looking closely at the needs and requirements of the population likely to be using these escape routes. You should also bear in mind the likely conditions under which these routes may be used. The emergency situation might be caused, or aggravated by such conditions as: severe weather conditions, flooding, loss of power and light, rioting mobs, overcrowding, or pollution.

The emergency evacuation checklist should again prove useful here because it poses questions about the desirable characteristics and features which need to be avoided or addressed. The list of potential difficulties is not necessarily exhaustive; thus, you should try to adopt an open and inquiring mind while conducting the inspection prior to selection and during any subsequent reviews. Look at the situation from the perspective of prospective evacuees, envisioning the conditions under which they may be literally running for their lives – understanding that those escaping may not be in the most cautious and deliberate of moods. Panic tends to preclude clear thinking and often suppresses rational behavior; so endeavor to make their journey to safety as easy and secure as possible.

Often the obvious, or the desirable, path to safety will make use of passageways and pathways which are not in regular use, at least not as pedestrian pathways. Because they are not used as paths, you have to ensure that they will always be open and free to use. I have often come across exit and escape routes which have been compromised by someone using them for alternate purposes such as storage areas or rubbish dumps. Usually, because it is unauthorized, this use has been overlooked by those responsible for health and safety or security. Too often, the clearly marked emergency exits, particularly in the loading area of a business, are locked for "*security*" reasons to prevent theft. Recently, I read about a case in which the fire exit in a theatre was locked to prevent it being used for unpaid entry or the area being used by the homeless for sleeping accommodation. Allegedly, this situation was not recognized by management until there was a fire in which several people died.

**If your population includes, or is likely to include, elderly or disabled persons, then you must pay particular attention to their needs and consider the use of hand rails or other aids...**

Other problems to watch out for include the possibility of parked vehicles causing an obstruction, locked gates or narrow passageways impairing the escape, rough surfaces causing people to stumble, and steep or slippery slopes which may also be the cause of accidents.

If your population includes, or is likely to include, elderly or disabled persons, then you must pay particular attention to their needs and consider the use of hand rails or other aids which might assist such people in their flight.

It is a good idea to conduct your escape route survey in collaboration with whoever is responsible for security or health and safety so that, between you, you will be able to spot and recognize the various good and bad points relating to each of the possible pathways.

The final selection will be a relatively simple matter of choosing those with the fewest snags or those which are the simplest to follow or are the easiest to maintain.

Your report, together with your recommendations, should be based upon the information which you noted as you followed the paths and consulted the checklist. Retain all the evidence of your investigation – it could prove to be very useful at some later date.

## 4.2 Concerns for the Disabled

Today, it is generally accepted and often required that we should give equal concern and opportunity to those who are disabled. Thus, we commonly find disabled persons as a normal part of our community, which means they will be in our premises and need to be considered whenever we are thinking about planning for emergency evacuation.

In the past, a common approach to the disabled evacuation question was to look at methods of so-called invacuation, in which the less mobile sought, or were offered, refuge in a relatively safe space within the building.

Because such an invacuation strategy reduces the need to provide complex evacuation arrangements, it brings reduced costs in the event that a safe refuge could be found or made available. Indeed, this approach is still considered by many to be a satisfactory solution to a complex problem. Others would see this type of option as, at best, a temporary measure, one which can provide a breathing space for the less able while they are following an exit or escape route along with everybody else.

The major problem with invacuation is that we might appear to be abandoning our disabled fellows in the interests of the majority, who are hoping to reach safety as quickly as possible. Such actions imply little regard for those who are less fortunate, treating the disabled as second-class citizens who, in the worst case scenario, could be at risk of being abandoned to their fate within a crumbling building.

Invacuation should not be confused with the practice known as *shelter in situ* or *shelter-in-place*, conditions in which the safest place to take refuge or cover from an actual or perceived danger is the person's current location, such as the person's own home, business place, or school. Such shelter equates to the "go in, stay in, and tune in" advice developed by the UK's independent National Steering Committee on Warning and Informing the Public as being the best general advice to give people caught up in most emergencies. It is the best advice for those instances in which the individuals concerned are completely unprepared and the incident is a large-scale threat to the environment rather than an emergency situation in a specific building or location.

A copy of the full text of the advice from this advisory body is included in "Evacuation and Shelter Guidance" in the EEP Toolkit which accompanies this book. Another good reference source in this connection is British Standard BS 8300:2009 Code of Practice, Design of buildings and their approaches to meet the needs of disabled people. This document updates the original 2001 version which superseded the earlier code which formed part 8 of BS5588.

I also recommend that you take a look at BD2466, which is in the EEP Toolkit.

In 1990, the United States Congress passed the Americans with Disabilities Act (ADA), generally considered the flagship piece of legislation on the subject of accessibility. Following on from this, the National Fire Protection Association (NFPA) is the primary source of information in connection with evacuation planning for the disabled, and in March 2007, the NFPA published *Planning Guide for People with Disabilities*, which is also included in the EEP Toolkit.

Arrangements for the evacuation of disabled persons should be compatible with the general evacuation strategy and the emergency fire action plan for the premises in question. The responsibility for implementing the plan and evacuating such persons safely in the event of an emergency will rest with those who are deemed to be responsible for the care, health, and safety of those occupying, visiting, or passing by the premises in question. An evacuation plan for disabled persons must be in place as part of the existing emergency plans and should not rely upon the intervention of the Fire and Rescue Service to make it work.

That doesn't mean that the firefighters in attendance will refuse to help the disabled, but you should not consider it part of their responsibility. Your plans must be self-sufficient in that sense.

The precise legal implications will vary from jurisdiction to jurisdiction, but the principle holds true throughout the world that it is morally wrong to rely upon others when a little forethought and planning can provide a built-in solution which can be tested and proven. This planning is especially necessary when those others are based some distance away from the location in question and their response is on a "best efforts" basis. In the time of need, outside emergency workers may be overwhelmed with other calls or simply delayed, whereas the occupants of the building will already be present when the alarm goes off.

## 4.2.1 Elevators

Purpose-designed evacuation elevators and fire-fighting elevators have features and safeguards which may allow their use in the event of fire. Other elevators are not normally considered suitable for fire evacuation purposes. Therefore, in your planning, you should expect all elevators to be deactivated whenever the fire alarm goes off.

**If you are at all uncertain about the availability of enough trained staff, don't include unprotected elevators or escalators as elements within any emergency evacuation plan.**

It may be possible to arrange for certain elevators to be on a timer to permit them to remain in service for a short while after the alarm has been raised in order to allow for the evacuation of disabled people. Obviously, this concept has to be discussed and agreed upon with both the elevator engineers and the fire and rescue professionals before it can be implemented as an effective evacuation measure. Furthermore, you will need to have an education and awareness program to support the concept in place before it can be said to be fully implemented and generally regarded as a reliable exit mechanism.

If you are at all uncertain about the availability of enough trained staff, don't include unprotected elevators or escalators as elements within any emergency evacuation plan. Use of elevators and escalators – in the absence of well-trained staff – may introduce additional hazards in an emergency situation such as a fire.

In two major hotel fires in the United States, which are often quoted as reference cases, a number of occupants died when elevators in which they were riding descended to the fire floor and the doors opened automatically, thus exposing the passengers to intense heat and fumes. Other occupants had managed to escape earlier by using the elevators, but the situation had rapidly changed for the worse. The first of these two unfortunate events occurred at a fire in the MGM Grand in Las Vegas, Nevada, in 1980, in which five people died attempting to escape in an elevator. In 1986, at the Dupont Plaza, San Juan, Puerto Rico, three people died in a similar manner. Of course, these two events led to some serious investigations and reviews of the procedures, regulations, and recommendations regarding the use of elevators in an emergency.

During a fire, once the fire and rescue service is in attendance, those professionals may operate an elevator override system in order to use an elevator to access the fire. As a result, all elevators in the building for use by the fire and rescue service may return automatically to the fire service access level where they will park. Once this happens, it is not possible for members of staff or the public to call elevators as they will be under the control of the fire and rescue service. Fire-fighting elevators may therefore only be used in the early stage of the evacuation process in full agreement with the fire and rescue service.

When elevators are used for evacuation, disabled people should be directed to make their way to the elevator point and to use the communication system to contact the persons coordinating the elevators, usually security personnel, so that security personnel are aware of the floor where the disabled are waiting. Also, there may be a communication point through which the coordinator can be contacted by any disabled persons in a temporary waiting space adjacent to the elevator. (BS 5588: Part 8 contains information on evacuation elevators).

### 4.2.2 Fire Compartmentalization

When buildings are divided into separate fire compartments, you may find it possible to evacuate people horizontally through the building into another fire compartment and away from the emergency situation. Where such evacuation is not available on the affected floor, it may be available on a lower floor, which may prove more acceptable than traveling all the way to a ground floor designated exit point.

## 4.2.3 Temporary Waiting Space or Refuge

Disabled people should not always be held back to wait for the main flow of people to be complete before they begin their departure. If people need to rest or if they feel uncomfortable with people behind them, you may find it appropriate to design a plan that allows resting in temporary waiting spaces provided along the exit route. Such a temporary waiting space (often described as a refuge) is an area that is separated from a fire by fire-resistant construction and provided with a safe exit route, thus constituting a temporarily safe space for disabled persons. A temporary waiting space can be used as a safe resting place as well as a place to wait during evacuation until the necessity for a full evacuation is established.

Some cautions about the use of such temporary waiting spaces include:

- Such rest stops should not obstruct other people leaving the building; it may be safer for all concerned if the slower moving people follow on behind the main flow.
- Leaving the disabled in a temporary waiting space simply to await the arrival of the fire and rescue service is tantamount to abandoning them to their fate. It also places too much reliance on the firefighters to reach them early enough and have the time and resources to move these people to safety, and it can divert their attention from other more pressing duties.
- A temporary waiting space needs to be equipped with some type of communications. The communications device should be of the simplest and most reliable kind. In the event of a major emergency, sophisticated digital communication systems such as voice over internet protocol (VOIP), or the local cell phone network may become overloaded, damaged, or unavailable. Stick with traditional analog communications; they are more dependable under the difficult circumstances usually associated with the need for an emergency evacuation, such as a fire, flood, or power failure.
- Space may be restricted. Many temporary waiting spaces within stairways can accommodate only one wheelchair at a time. Thus, a suitable evacuation strategy needs to be in place when you have more than one wheelchair user. In such cases, as one person progresses on his or her journey, the next person can take his or her place in the temporary waiting space.

## 4.2.4 Managing Reasonable Adjustments

An organization responsible for managing premises that are used regularly by disabled persons will be expected to make "reasonable adjustments" to the environment and the facilities available in order to cater to the needs of a wide variety of persons who use those premises, including the disabled. Failure to make allowances for these needs is discriminatory negligence in the eyes of the law, and, in most parts of the world, penalties are attached to this type of failure. Questions of corporate citizenship and issues to do with brand and image also need to be taken into account. Obviously, the term "reasonable adjustments" is subject to interpretation that would ultimately take place in a court of law. The law court is somewhere we would all wish to avoid; thus, it behooves us to do our best to anticipate the needs of our fellow citizens, especially under extreme or emergency conditions.

**Disabled people may be more willing to facilitate their own evacuation when they know that this is not going to be required of them during a practice exercise or for a known false alarm.**

Disabled people are individuals with particular needs, and each person should be treated accordingly. When developing a plan, you should avoid the tendency to over-play the safety issue to the detriment of the independence and dignity of disabled people. What disabled people are prepared to do in exceptional circumstances may differ significantly from what they should reasonably manage in their everyday activities, especially if basic, reasonable adjustments, such as those suggested below, have been made. Evacuation plans for disabled people should be prepared with the view of what is required for "the real thing" and should take into account what is practical and achievable in exceptional circumstances. This level of effort required of a disabled person may not be acceptable for a practice exercise or known false alarm or in everyday activities. Thus, in a true emergency, it's very possible that solutions could be used that may not be appropriate in most circumstances.

Disabled people may be more willing to facilitate their own evacuation when they know that this is not going to be required of them during a practice exercise or for a known false alarm. Some disabled people are put at a risk of injury when being carried. It is therefore necessary that in these cases, the evacuation policy should include a method of minimizing the need to evacuate for known false alarms, drills, or exercises.

Some disabled people may have multiple impairments, and their needs may be distinctly different. For example, a person with dual sensory impairment (deaf and blind) may have needs quite distinct from those with a visual impairment or hearing impairment alone.

## 4.2.5 Mobility Impaired People

A wide range of people fit into the category of "mobility impaired." Issues relating to this group of people may also be relevant to those who have asthma or heart conditions. The preferred options for evacuation by people with mobility impairment are horizontal evacuation to the outside of the building or into another fire compartment; or vertical evacuation by means of an evacuation elevator, eventually arriving at a place of safety outside the building.

It may not be possible to evacuate the disabled within a short period of time. Many of them may be able to manage stairs and, perhaps, walk longer distances, especially if short rest periods are built into the evacuation procedure. Reasonable adjustments such as suitable handrails may be of great benefit, though some people will be able to use a handrail only if it is on the side appropriate to them. Some individuals may prefer to shuffle or slide down the stairs after the main flow of people.

Several types of specially designed mechanical equipment are available to move people up or down stairs; however, timing and obstruction of exit or escape routes for others are prime considerations when considering the use of this type of equipment for emergency evacuation purposes.

Wheelchair users normally have difficulty evacuating the premises when they are on a floor above or below the regular accessibility level (normally the ground floor). However, some frequent wheelchair users may be able to walk slightly and therefore be able to assist with their own evacuation or even achieve independent evacuation. Assumptions should not be made about the abilities of wheelchair users; they should be consulted about their preference.

If horizontal evacuation or the use of an elevator is not available, the individual may require assistance from one or more people. It may even be necessary to carry the person down (or up) an exit or escape stairway.

**Carry-Down Techniques:** Your staff needs to be aware that a number of methods may be appropriate for carrying people with mobility impairment down stairs using two, three, or four people. All carry-down techniques require a risk assessment (including a manual handling risk assessment) as well as training and practice for the carriers. Important issues to be considered are:

- "Will enough physically capable carriers be available?"
- "Is the width of the stairway sufficient for all of the team of carriers to move freely and safely."

**Evacuation chair:** A typical evacuation chair looks somewhat like a deck chair with skis and wheels underneath, and it is maneuvered by one or two people. When placed on the stairway, it slides down the stairs. The wheels at the back facilitate movement on the flat, but these chairs are not suitable for long distances. Disabled people may not feel confident using these chairs, and it is not always possible for wheelchair users to transfer into an evacuation chair or to maintain a sitting position once seated in one. Therefore, evacuation chairs should not be considered as the automatic solution to the evacuation requirements of all disabled persons.

The provision of evacuation chairs needs to be accompanied by a full system of evacuation for disabled people, complete with trained operators who are familiar with the equipment. Make sure that regular practice takes place. In most instances, practice should not include the disabled person, although some may wish to practice being moved in the evacuation chair. It is more appropriate for those people who are trained to use the evacuation chair to play the role of the disabled person during practice; this will increase their confidence in using the equipment.

In one instance, during an exercise at a university, a disabled member of staff took a careful look at the evacuation chair which had been provided for her and came to the conclusion that she would probably suffer severe damage if she were carried in such a device. She suffered from severe *osteogenesis imperfecta*, otherwise known as brittle bone disease, and she estimated that it would take her about two and a half years to recover from the damage that would be caused by being carried down two flights of stairs in such a device. Naturally enough, she did not participate fully in the evacuation exercise, and her hypothesis was not fully tested. Another solution, involving horizontal evacuation using her own wheelchair and a special elevator, had to be found for her. However, this alternative couldn't be arranged at short notice – but the incident did highlight the value of a realistic exercise in proving and improving the emergency evacuation plan.

Where emergency evacuation devices are used, they may be allocated to particular individuals with specific needs and either kept alongside their work location or in the most suitable temporary storage space close to them.

**In buildings where there is open public access, it may be advisable to provide special evacuation chairs at suitable points within the building.**

In the event that a visitor requests this method of evacuation and a device is available together with suitable people to assist, whoever is responsible for booking the visitor into the building should arrange for the device to be brought to the most suitable point of the building for the duration of the visitor's stay.

In buildings where there is open public access, it may be advisable to provide special evacuation chairs at suitable points within the building. One on each staircase at each level may be an expensive option. However, provision of these on the top floor of the building with a system which allows them to be brought immediately to the temporary waiting space may be an acceptable solution depending upon the fire safety measures in place and the circumstances of the case.

**Wheelchair evacuation:** It may be possible to move a person down a stairway in a number of ways, while the person remains in the wheelchair. The person can be carried down by two, three, or four persons holding the wheelchair at one of the rigid points in each corner of the wheelchair. The team then lifts and moves the wheelchair and its occupant up or down the stairs. The person should be moved after the main flow of people. (Motorized or other heavy wheelchairs are not suitable for this purpose.)

With some wheelchairs, it is possible to tilt the chair on its axis so that it is virtually weightless on the stair. With either one or two people holding onto the chair by a fixed point at the rear of the chair, the wheelchair can be maneuvered allowing the weight of the person to take the chair down the stairs. A minority of wheelchair users are able to make this "wheelies" maneuver unaided, but it is generally only practical on a short flight of stairs. Where it is considered practical, individuals should be consulted about their willingness to use this method, and staff should be trained to assist them in these maneuvers. In all such scenarios, it is important that where possible the dignity of the person should be maintained throughout, as this consideration is likely to enhance a swift egress. Instruct your staff to ask rather than assume.

People who use electrically powered wheelchairs may have less mobility than those who use manual chairs. There will be exceptions to this rule, so it is important to consult with the disabled person. If there is no suitable elevator to facilitate a person's evacuation, then the wheelchair may need to be left behind in the building because of its weight and size. This will mean that some other method of carrying the person down (or up) stairs will be required, which may involve special equipment such as an evacuation chair. Persons with restricted mobility may require more assistance when leaving the building, assistance which may be difficult if the person is heavy. You should make sure that your staff is fully trained in the best methods for moving people safely without harm to themselves or the person being moved.

**Other chairs:** Carrying someone down (or up) a flight of stairs using an office chair can be achieved in a similar way as when using a wheelchair. Any stout stable office chair can be used for this purpose, although one with armrests is preferable.

**Alternative Methods:** Some people may find it difficult or impossible to transfer from their wheelchair into an evacuation chair or other evacuation aid; a hoist may be required to assist with this movement. The process can be difficult and requires suitable training. In these cases, a risk assessment of the use of elevators within the building for evacuation purposes may reveal that this solution presents less of a risk. Where the use of elevators is deemed appropriate, consideration could also be given to the location of workstations or points of service used by disabled people to facilitate their egress.

## 4.2.6 Hearing Impaired and Deaf People

You should not assume automatically that a hearing impaired person cannot discern the operation of the fire alarm. However, some hearing impaired and deaf people will need to be alerted to the fact that an evacuation is in progress. Where the fire alarm has audible sounders or there is a voice alarm system, they may not be able to discern the alarm or any information being broadcast. If sound enhancement systems are provided within a building, it may be possible to transmit the message through that system, such as a hearing loop or radio paging receiver.

Flashing beacons and vibrating pagers are examples of reasonable adjustments, and either type can be used as part of the fire alarm system to alert hearing impaired people to an evacuation. Pagers can also be used to communicate with other people who are part of an assisted evacuation system. The pagers can be used to inform people that there is a need to escape and also tell them in which direction they should travel; however pagers cannot always be provided. Visual warning may not be appropriate in all buildings, for example where there are other lighting conflicts. Where aids are not available, then a suitable buddy system will be required and should be implemented on a formal basis. A buddy system may be the best method of alerting a hearing impaired person to the operation of the fire alarm.

Where other staff members are expected to alert hearing impaired or deaf people to the need to leave a building, you should make sure they are trained in deaf awareness. Often fire wardens or fire marshals sweep a building to ensure that no one is left on the floor. These people can, and should, be trained to look for signs that a hearing impaired person is present who may not have heard the alarm.

When checking, fire wardens or fire marshals should not rely on just a vocal call but should be trained to physically check all of the areas for which they are responsible.

Instruct your staff that a person who does not appear to react in a logical manner during an evacuation may not actually be aware of the alarm.

Shouting louder may not help. It may be necessary to explain what is happening with signs, a written note, or a pre-prepared short written instruction.

Some hearing impaired and deaf people do not use English as a first language. It is important that a plain English translation of the fire action is provided. It may also be advantageous for a pictogram to be provided to support the written information. Deaf people may prefer to have instructions explained to them using British Sign Language (BSL), American Sign Language (ASL), or whatever other sign language they are accustomed to.

You should take care to ensure that deaf or hearing impaired people who are working alone in a building know what is happening. In these instances, installing a visual alarm system or vibrating paging system may be required.

## 4.2.7 Visually Impaired and Blind People

You will need good signage and other orientation aids to assist in evacuating people who are visually impaired. Visually impaired people who have some sight may be able to use this during the evacuation in order to make their own way out of the building as part of the general exodus. Where the physical circumstances are appropriate, they may have no problems leaving a building.

You will find that existing design elements within the building may help visually impaired persons to cope with their own evacuation unaided. Useful features of good building design, with reasonable adjustments, include good color contrast; handrails on escape stairs; contrast to the nosings on the stair treads; markings on escape stairs; color contrasted or different texture floor coverings on exit and escape routes; or other directional information. Orientation aids, tactile information, and audible signs may further reduce the need for assistance. Where audible signals are used, any potential interference by the fire alarm operation needs to be considered. Good color definition and accessible signage will help visually impaired people to find their way through a building. If you are able to extend these systems to include the exit and escape routes, you may be able to reduce the need for assisted evacuation.

Improving circulation and orientation within the premises can prove to be of great benefit.When you provide logical, clearly marked routes to the exit or escape stairs, it will not only assist visually impaired people but will also benefit all the other users of the building – and, in the daily use of the building, will make life easier for visitors and infrequent users of the building. Where there is a lack of orientation information, staff assistance may be necessary to provide guidance out of the building.

A person with dual sensory impairment may require a very different approach to that from someone who is blind because of the need to consider additional communication.

Wherever people are normally accompanied and assisted by a guide dog, they may prefer for the dog to assist them on their way out of the building. Others will prefer to take the responsibility away from the dog and request a human assistant. In these cases, you will need to allocate a buddy to the person, as well as providing a person to look after the dog. Consideration should be given to the welfare and comfort of the dog, arranging for the animal to be fed and watered if the evacuation lasts for a while.

Visually impaired persons may not be able to locate exit signs easily. They may not be aware of the travel direction to get out of the building, although they may be able to recreate their way out along the route which they used to enter the building. If you use the exit or escape routes as part of the general circulation space within the building, you have a greater chance that visually impaired people will become more familiar with these routes and able to use them in the event of an evacuation.

Visually impaired people may not be able to read the fire action notices provided in most buildings, as these are often in small typefaces.

You should consider making instructions available in Braille, large print, or on audiotape.

It can be useful to provide a tactile map of the escape routes and to provide orientation training to any visually impaired staff so that they become more aware of the options for evacuation. Tactile maps and large print can be obtained through a number of organizations that provide accessibility information services. It may also be possible to produce large print in-house.

Check for stairs with open risers on exit or escape routes, since visually impaired people in particular may have difficulty on stairs which have open risers. Where these are present, make adaptations to the stairs to make them safer or arrange for assistance on the stairs for those who require it. Alternatively, a different stairway may be made available.

When any internal physical changes are made in a building, such as the construction of partitions or the rearrangement of office furniture, it is important that you make these changes known to any visually impaired people in the building.

### 4.2.8 People with Cognitive Impairment

Your staff needs to be aware that those with cognitive impairment often have problems comprehending what is happening in an evacuation situation, or they may not have the same perception of risk and danger as non-impaired people.

In the case of certain conditions such as dyslexia, dyspraxia, or autism, people may be seemingly unaware of their impairment. Many people with a learning disability also have other impairments. They may have mobility difficulty, impaired vision, or poor hearing. Those who have cognitive disabilities may move more slowly than the main flow of people; thus, you may need to provide for a slow and fast lane on the exit or escape stairways if the stair width allows for different rates of egress, not only for the handicapped but for anyone else in the building who needs to take his or her time on the stairs.

You may not always be able to tell that persons have an impairment which affects, or inhibits, their ability to orient themselves around the building. Staff should be made aware of this and taught to be tactful when assisting those who may seem lost or unsure of what to do during an evacuation.

The key elements of evacuation planning for people with such challenges are:

- Staff awareness.
- Having someone to help.
- Familiarity with the various routes of travel.
- Providing, or developing, an environment built to be simple to use.

The most effective form of assistance for a person with a cognitive impairment is to have a support worker or assistant there to help, but not every person in this category will be accompanied by a helper; thus, you should make sure that efforts are made to enable the disabled person to understand how to leave the building on his or her own rather than assuming that a helper will take on this role.

**Practice of the route options can reduce the requirement for staff assistance dramatically.**

You will find that no single system would be equally effective for all cognitive needs. Orientation information and color-coding of escape routes can be useful. Consistency of color is important, though not everybody will benefit from color-coding. It is also important to realize that some colors do not show up very well under some lighting conditions, especially when the lighting is reduced due to smoke or is simply dim. People may have difficulty in eliciting the right information from some signs (there is evidence that some people with cognitive impairment use both symbols and words on signs). They may need to have the evacuation plan read aloud and explained to them. A video or DVD explaining and demonstrating what to do in an emergency can also be an advantage as can a photographic explanation of the route. Signage may be only part of the process to help people with cognitive impairment find their way. Building features and building layout are also important. Use of escape routes for general circulation through the building is an advantage, as some people may be reluctant to take an unknown route from the building.

Practice of the route options can reduce the requirement for staff assistance dramatically. Practice is essential in all those cases in which assistance is required, especially in situations where one person is responsible for a number of others, in a classroom or workshop situation, for example. People with a learning difficulty or poor memory may need to practice their routes for escape frequently, perhaps on a monthly basis. If so, this need for regular practice should be written into a person's PEEP.

In any case, you have to make sure that all people with special needs who are likely to be in the building on a regular basis have their own PEEPs. Furthermore, they should be involved in the preparation and maintenance of their own plans. If at all possible, they take full responsibility for the whole development, delivery, and maintenance process, providing, of course, that they feel they have the capability and the desire to take on this level of responsibility for their own safety.

## 4.3 Planning for the Disabled

As we mentioned in the previous section, you will need to consider a number of different types of impairment when preparing plans for specific individuals or groups who have special needs, and we shall consider the approach to each of these categories in turn. These suggestions should be read in conjunction with the comments made in the previous section regarding "Concerns for the Disabled."

Throughout the process, you and your staff should always respect the dignity and independence of all those concerned when planning for their evacuation and bear in mind their willingness and capability to make a special effort in an emergency situation. Also, take account of their special needs, establishing what is possible and what is required. The whole process should be seen as planning with rather than planning for.

For the purpose of preparing PEEPs, we can recognize the following categories of disabled person:

- Mobility impairment.
- Hearing and deafness.
- Visual impairment and blindness.
- Cognitive impairment.

Persons in each category will have their own specific set of concerns to be taken into account and particular questions which need to be addressed when preparing to plan for their evacuation. Do not make assumptions on their behalf, especially not before engaging them in a dialog about their needs, capabilities, and willingness.

## 4.3.1 Plans for Mobility Impaired People

When writing a plan for, and with, someone who has impaired mobility, or who simply uses a wheelchair, the provision of reasonable modifications and changes together with the following areas of the environment should be taken into consideration:

- If stout handrails on all exit and escape routes are provided.
- Whether handrails are, or should be, on one or both sides.
- How far the person will be expected to travel on the routes in question.
- The availability of separate fire compartments to be used as temporary rest areas or to provide access to alternative escape routes.
- If special evacuation chairs, or chairlifts, can be provided.
- The location of any lift or elevator that can be used in the event of a fire.
- Whether willing and able people will be available to provide assistance.

**Interview Questions for Mobility Impaired People:** Be prepared to ask a number of questions when interviewing persons with impaired mobility regarding the development of their personal evacuation needs and expectations. A good starting point is to describe to them what is already in place for everyone and try to make sure that they do understand and appreciate the kinds of emergency you are planning for. Further questions may arise as the conversation develops, but the following questions should cover the main ground. All of these questions need to be prefaced with "in a real emergency" because it has to be made clear that we are not about to discuss routine or regular travel around the place.

In an emergency situation:

- Would you be able walk down the stairs, either with aid or unaided?
- How far would you be able to walk unaided?
- Could you shuffle or slide down stairs without any assistance?
- If you could shuffle or slide, how many flights of stairs could you manage, and would this be increased if assistance was made available?
- How many people do you think you would need to assist you?
- How many times might those assisting you need to stop for rest?

- Would handrails be of use to help your evacuation?
- Are there any places along the escape route where aids might assist you?
- How might your mobility be worsened, such as by smoke, etc.?
- Is your wheelchair electric powered or manual?
- Can you be carried in your wheelchair?
- Finally, would you be willing to take part in an emergency evacuation drill, or would you be prepared to evacuate only in the event of a real emergency?

## 4.3.2 Plans for Hearing Impaired People

When you are writing a plan with someone who has a hearing impairment or who is deaf, establish if any of the following are, or could be made, available:

- Some form of visual warning in the fire alarm system.
- A text-phone connected to the telephone network.
- A vibrating pager dedicated to emergency use.
- A fellow member of the team or department where the impaired person will normally be or someone who would take responsibility for alerting those with a hearing impairment.
- Fire wardens who would be responsible for ensuring everybody responds to the alarm.
- An appointed buddy for the individual.

**Interview Questions for Hearing Impaired People:** The following questions should be asked when you are interviewing persons with impaired hearing. These questions are designed to give a clear indication of their emergency requirements. Preface the questioning by describing what is already in place for everyone, and try to make sure that they do understand and appreciate the kinds of emergency you are planning for.

- Are you likely to be in the building out of hours?
- Are you ever likely to be alone in the building?
- Can you recognize the fire alarm wherever you are in the building?
- Do you work as part of a team or in a group environment?
- Do you have a dedicated text number?
- To what extent do you move around the building?

The answers to these questions should provide the right level of information upon which you can start preparing a suitable plan for this type of person.

## 4.3.3 Plans for Visually Impaired and Blind People

Visually impaired people in particular may have difficulty on stairs with open risers; these should be avoided on escape routes wherever possible. If an alternative stairway is not available, you may need to arrange for either assistance or for the stairs to be adapted to make them safer.

Whenever any internal physical changes are made in a building, such as the erection of partitions or the rearrangement of office furniture, it is important that these changes are made known to all visually impaired people who use the building.

When approaching the task of writing a plan with someone who has a visual impairment, the following information should be considered and taken into account:

- The type of fire alarm system which is currently installed and whether it can be modified or upgraded to accommodate the needs of the visually impaired.
- Suitable marking of all exit and escape routes.
- Availability and awareness of orientation information.
- Fire instructions in accessible formats.
- Clearly visible step edge markings on all exit routes, escape routes, and stairways.
- Stout handrails provided on all escape routes and stairs.
- Whether the stairways have open risers.
- Whether external escape routes cross open areas where people might lose their way.

**Interview Questions for Visually Impaired and Blind People:** Ask the following questions when interviewing a person with impaired vision; they are designed to give a clear indication of the person's emergency requirements. As when you question other disabled people, preface the questioning by describing what is already in place for everyone and try to make sure that the person does understand and appreciate the kinds of emergency we are planning for:

- Are there times when you work alone in the building?
- Do you occasionally work out of hours?
- Are you aware of the positions of all exit routes, exits, and escape paths?
- Can you use the exit routes and escape paths unaided?

- Do you work as part of a team or in a group environment?
- To what extent do you move around the building?
- Do you normally have a guide dog with you when you are in the building?
- If so, would you want your guide dog to lead the way in an emergency evacuation, or would you prefer to have someone help you both out of the building?
- Can you read the evacuation instructions? If not what format do you need them in?

## 4.3.4 Plans for People with Cognitive Impairment

When writing a plan with persons who have a cognitive impairment, start by determining what they understand and develop the plan based on how they will find the escape routes and what reasonable adjustments they may require. Consider:

- The type of fire alarm system which is available.
- Marking of the alternative exit routes, exits, and escape routes.
- The availability of suitably clear orientation information.
- If fire instructions are in formats which they will understand.
- Step edge markings on the exit and escape stairs.
- Handrails on the exit and escape stairs.
- The need for two-speed traffic on the stairways and whether the stairs are wide enough to allow this.
- Whether there are suitable rest points where others may pass while the disabled person and the assistant(s) pause.
- If stair risers are open.
- Whether there are external open escape routes with the possibility of people losing their way.

**Interview Questions for People with Cognitive Impairment:** Ask the following questions when interviewing persons with cognitive impairment; they are designed to give a clear indication of people's emergency requirements. As with questioning other disabled people, preface the questioning by describing what is already in place for everyone and take particular care to ensure that they understand and appreciate the kinds of emergency you are planning for.

- Do you ever work alone in the building?
- Do you sometimes work out of hours?

- Do you know what the fire alarm sounds like?
- When you hear the fire alarm, do you know what to do and where to go?
- Do you work as part of a team or in a group environment?
- Are you likely to move around the building?
- Can you read the escape instructions? Do you understand them? If not, what format would you need them in?

Once you are armed with the information gathered as a result of considering what is in place, or could be made available, together with interviewing the people concerned, proceed with developing suitable PEEPs for these various categories of disabled people. Remember to engage them in the process as much as is practical. Their engagement ensures that they get the most benefit from the process and enhances their self-respect as well as their confidence. These factors all contribute to the successful employment of the associated escape procedures.

The next chapter deals with the development process in more detail. It also covers the development and distribution of PEEPs for other classes of people who may need their own PEEP.

**People with special requirements...may experience particular difficulties when following the regular emergency evacuation procedures because of a disability of some sort.**

## 4.4 Personal Emergency Egress or Escape Plans (PEEPs)

PEEPs are the customized emergency evacuation plans you will provide for people who have specialized, or unique, evacuation needs. Three broad categories of people need to be considered in this context:

- Those who may have special requirements in regard to the evacuation procedure.
- Those who may have variable requirements in regard to evacuation.
- Those who may have temporary or short-term requirements.

**Special requirements:** People with special requirements would typically be those who may experience particular difficulties when following the regular emergency evacuation procedures because of a disability of some sort. Some of them will require the help or assistance of others, some may require access to special equipment, and some may need to use alternate or modified evacuation routes.

**Variable requirements:** People with variable requirements would be those whose routines or work patterns takes them regularly to different locations; thus, they may need to escape from any one of those locations, each of which may have a different emergency procedure and subsequent assembly area. For example, university students may attend lectures routinely in a number of different buildings throughout the campus, or maintenance workers may expect to be operating anywhere within the properties which they maintain.

**Short-term requirements:** People with short-term requirements would be those who are only temporary or casual occupants and thus will not be engaged in the regular training and awareness program. They will require a guide to evacuation which they can carry around with them. In the event of an emergency or a drill , they will then be able to follow the directions in their PEEP to exit and escape from the building.

The following principles should be borne in mind when developing, implementing and maintaining a PEEP:

- The same rules of courtesy and respect should be applied to both disabled and non-disabled people alike. Disabled people should not be treated as some kind of inconvenient "health and safety" problem which needs to be resolved.
- Look at, and work with, the person, not his or her disability. Needs and preferences will vary widely between individuals.
- Disabled people should be involved meaningfully at all stages in the development, implementation, and review of their PEEPs. They may even prefer to manage this whole process for themselves; in that case, you must ensure that they are given the right training and provided with suitable guidelines or templates to complete the task.
- In any evacuation situation, do not make any assumptions about the type and degree of assistance a person might need; ask the person. The individual disabled person best understands the nature of his or her impairment.

### 4.4.1 Communication and Training

Good communication and appropriate training for all levels of staff and management regarding the fire or emergency evacuation process are vital to ensure success. It is important for staff and managers to understand fully the evacuation plan and the fire safety strategy for the building so that they are able to render maximum assistance to disabled persons irrespective of the nature of their impairment. Once you provide a fully integrated PEEP system, safety will improve for everyone using the building and any weaknesses in the existing evacuation plans will be identified.

All persons who are involved in the process of providing evacuation plans should have a good standard of equality awareness and disability awareness to ensure that they do not discriminate inadvertently against disabled people. The additional training you need to provide will depend on the role of each individual but may include:

- Disability awareness.
- Disability evacuation etiquette.
- Moving, lifting, and handling techniques.
- Appreciation of fire-resistant enclosures.
- Good practice in communication, including the use of communication systems.
- The importance of using pre-planned routes.

Staff members have a vital role in communicating the evacuation plan to any disabled visitors, thus making the attitude and awareness of staff important. Recently, I was in a public building in which a notice was displayed on the reception desk, politely asking visitors to inform their host about any condition which might restrict their ability to use the stairs in an emergency. This was in a two-story building with no lifts or elevators. In other buildings, the message might need to be modified to include other potential areas of difficulty for visitors.

You should also provide information about the existing evacuation plans and arrangements within staff handbooks. One of my clients, Chartered Institute of Environmental Health, has copies of the fire escape procedure available as handouts for visitors to the head office. I have come across similar information sheets available in the reception areas of other buildings. Make sure that all those people who use a building on a regular basis are aware of the fire safety strategy and the measures which are in place. For example, if the building has fire compartmentalization to allow horizontal evacuation into another fire compartment, people operating the plan should take account of this. Staff members who might be involved in using the evacuation plan should fully understand their role (including where the function is out-sourced). They should also be made to feel confident in their skills, and the disabled people should feel that they can trust the process and all those involved in it.

Where staff members have specific roles, it is important that you allocate each role to at least one other suitably cross-trained person who will be available to cover for a staff member who is absent for some reason.

All systems and equipment used for evacuation purposes, such as pagers, will need a system of regular checks or testing to be in place. Otherwise, items are likely to fail, go missing, or get misplaced; and systems are likely to change, evolve, or become obsolescent, thus jeopardizing someone's safety.

## 4.4.2 Tailoring Plans to Suit Individual Needs

Each and every disabled person who is likely to be in any building on a regular or frequent basis will require an individual PEEP to be written and tested. This is especially true if the building has not been specifically designed and built for people with impairments. Engage the disabled person fully in the whole process right from the very start. This is not just a matter of courtesy; it is the only way to ensure that each individual has faith in the process and its outcomes. Individual commitment and input will result in a plan that is likely to be best practice for that person rather than simply one that complies with the rules and regulations.

Those who require PEEPs due to the nature of their activities rather than because of their disabilities should be able to develop their own plans based on the information available to them. Make a plan template available to them as a starting point. However, make it clear that they are responsible for the development and updating of these plans, which should also be tested from time to time, perhaps in combination with a fire drill or other emergency practice session such as a business continuity exercise.

## 4.4.3 People with Special Requirements

When writing plans for people who have special requirements, you are generally dealing with unique requirements tailored to the needs of the individuals, dictated by their restricted or limited abilities to deal with an emergency. Broadly speaking, these are people with some form of disability or impairment. However, it is perfectly possible that we may need to accommodate people whose belief systems or personal codes of conduct affect the way in which they wish to respond, or be treated, in an unexpected or dangerous situation. It is important to bear this possibility in mind when approaching the question of evacuation procedures at the individual level. You and your staff should try to avoid pre-judging the issues which may affect an individual's needs and requirements in this connection.

Many people who are disabled are normally accompanied by a guide dog whose primary role is to assist them in some way or another. Usually the bond between the human being and the canine companion is of a deep and affectionate nature; there is an emotional attachment which underpins their working relationship and supports the trust which is an essential element of the partnership. In an emergency, the welfare of the animal will be a major concern to the person the animal cares for. You should regard the individual and the dog as a close-knit team, probably self-contained and self-sufficient. Planning for them needs to take into account the needs and welfare of the animal as well as its duty and capabilities. You need to discuss all of this with the person concerned in order to determine the most appropriate course of action in an emergency for the benefit of both of them.

Because each of the plans written for this type of individual are entirely made to order, you need to adopt a flexible approach, taking into account all of the possibilities. The result will be a unique document, perhaps requiring some specialized equipment, probably with a need for assistance which will involve practice or training and alternative evacuation arrangements.

A checklist of aspects to be covered appears in the EEP Toolkit as a PEEP Checklist.

## 4.4.4 People with Variable Requirements

People with variable requirements are those whose routine involves activity in a flexible environment or variable location. For example, you might be dealing with college or university students who attend lectures and study in different buildings throughout the campus. Another group might be maintenance engineers who have the entire building as their normal place of work. Some of this type of work may be contracted out to third parties who are in regular attendance in various areas of the premises. In an emergency situation, all these people will need to have, and to follow, an evacuation plan which takes their lifestyle, work pattern, or study routine into account.

Plans for these people can probably be developed from a common template with a few options regarding the details of the exit and escape routes to be used. Ideally, they should take the responsibility for the development, testing and updating of their plans. However if they fail to deliver on this, you will have to direct someone to do it for them. Otherwise, a successful outcome of the whole evacuation planning process is in jeopardy.

## 4.4.5 People with Short-Term Requirements

In many places, you will identify those who are in attendance only for relatively short periods of time during which they may not become totally familiar with the working environment and may not be recognized as regular members of a particular team. For example, consultants might be brought in for a special project, and while they may be quite involved in certain aspects of what happens on the premises, they have no direct interest in regular routines like fire drills and business continuity exercises. Others who may have short-term assignments are auditors, inspectors, accountants, and advisers. As a visiting lecturer at Coventry University, I have to familiarize myself with the evacuation procedure for whichever building I happen to be visiting every time I go there. It is not a major chore, but it is a part of my routine whenever I go to work on a project or assignment in an unfamiliar location.

The easiest way to ensure that these short-term attendees are properly prepared for an emergency is for you to provide them with a simple, generic PEEP which gives them a quick, easy to comprehend overview of the emergency evacuation procedures. Clearly, it helps if the layout of the premises lends itself to straightforward exit and escape routes. Another useful attribute would be clear and consistent signage; in most scenarios, such signage is obligatory under the relevant fire regulations, but the execution of such a requirement is not always carried out to best effect. We have all seen instances of fire exit signs which could be easy to miss or misinterpret. On one occasion last year, my wife and I found it difficult to work out whether the fire exit signs were suggesting we should make our way along the corridor, up the stairs, or turn around and go back in the direction we had come from. As it happens, we were trying to understand from only a purely academic point of view, and I like to think we would have found our way out to safety if there had been a need – although I have no confidence that anyone's reading comprehension is any better in an emergency situation.

## 4.4.6 Format for PEEPs

Generally speaking, a PEEP will need to be something which can be slipped easily into a pocket or handbag to ensure that the individual concerned does carry it personally for reference purposes. I am an advocate of the tri-fold layout because it is easy to produce, easy for most people to handle, relatively convenient to carry and, perhaps, too big to be mislaid or lost.

Normally, I print out this type of document on light card stock or heavy paper to make sure that it remains firm and easy to handle. Otherwise, I find that the documents need to be replaced fairly frequently because they get rather tattered if someone carries them around as intended.

Here is an example of a tri-fold plan with the fold lines marked to make it easier to fold the plan correctly. This one includes a graphic diagram of both the exit route to get out of the building and the escape route leading towards safety and the assembly area. A written description of the exit and escape routes accompanies the drawings. The original template is included in the EEP Toolkit.

You can produce a PEEP in almost any format as long as the resulting document is practical. In other words, the user will want something which is fairly portable, easy to understand, and reasonably sturdy.

Sometimes it might be appropriate to provide users with a laminated version which might be easier for them to use and it will be fairly weatherproof.

**Emergency Exit Route**
Take the stairs to the Ground Floor and make for the main doors opposite the bottom of the stairway.
Leave the building and move away from the doorway so other people can get out and the Fire and Rescue services can gain access with their equipment

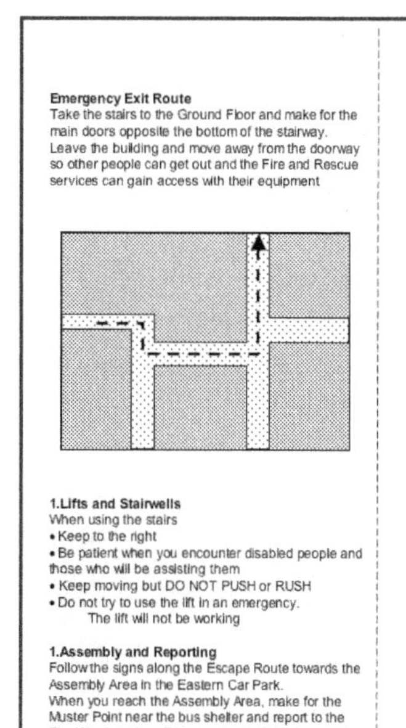

**1.Lifts and Stairwells**
When using the stairs
- Keep to the right
- Be patient when you encounter disabled people and those who will be assisting them
- Keep moving but DO NOT PUSH or RUSH
- Do not try to use the lift in an emergency.
  The lift will not be working

**1.Assembly and Reporting**
Follow the signs along the Escape Route towards the Assembly Area in the Eastern Car Park.
When you reach the Assembly Area, make for the Muster Point near the bus shelter and report to the Fire Warden who will be wearing a Yellow cap and vest, standing on a plinth.
Make sure the Duty Fire Warden records your name correctly on the Evacuation Register.

**Evacuation Procedure**
Whenever the Alarm Bell sounds continuously make your way from the Work Area to the Stairway and follow the signed Exit Route to the Exit Point, then follow the Escape Rout to the Assembly Area.

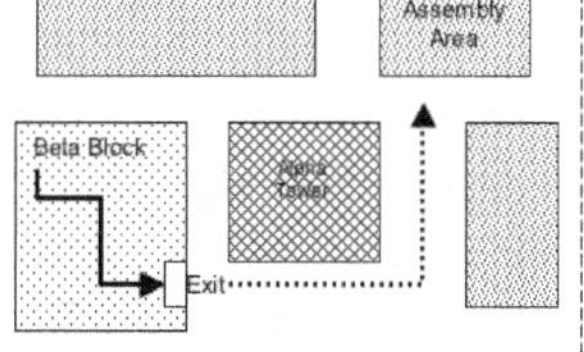


**Raising an Alarm**
If you spot a fire or smell smoke do not hesitate to raise the alarm.

- Fire Alarms are located inside the entrance door to each office area and at regular intervals throughout the corridors

If you have cause to believe that there is a non-fire emergency call Security on Extn 9999

- They will raise the alarm if it is necessary.

**Fire Fighting**
In the event of a fire DO NOT attempt to put it out.

This is a task which is best left to trained people who know which extinguishers to use and how to make effective use of them. There are a number of trained fire wardens stationed on every floor throughout the building.

**Personal Emergency Evacuation Plan for Bill Bailey**

**1.Test and Update**

**Test and Update**

| Version | Tested | OK? | Amended |
|---|---|---|---|
| Draft v.1 | 1st April 2010 | ✓ | 26th April 2010 |
| Final v.1 | 7th May 2010 | ? | 15th May 2010 |

**Introduction**
- This tri-fold PEEP is designed to guide the user to reach a place of safety (Assembly Area) when evacuating the Beta Block building

- Other personnel who are regularly working in the building will be trained in the use of exit and escape routes – they will not have a PEEP.

The Fire Marshall, identified by a red safety helmet and yellow vest, will take charge in an emergency. In the interests of everybody's safety his instructions take precedent over any other considerations or messages.

**Alarms**
The Alarm Bell will signal the need for the immediate evacuation of the building.
- An intermittent 2 second ring is an Alert or a Test.
- In the event of an Alert, get prepared to leave the building, ensuring you have your valuables and street clothes with you
- In the event of a Test, there will be a prior announcement on the Public Address system.
- Do nothing except remind yourself of the exit procedure and the route you would have taken.

**Return to Work**
When the emergency, drill or rehearsal is over the Duty Fire Warden will make an announcement.

The essential content of message will either be:

-"You may now return to work"
on which case you should make your way back to your desk
-"Report for work tomorrow morning"
on which case you should make your way home and report for work as normal on the following day
-"Go home and await instructions"
on which case it is likely that the building will remain closed for some while and someone from HR will contact you to tell you what arrangements have been made for you to continue with your work.

**Updates**
In the event of building closure someone from the HR department will make contact with you to keep you informed of any developments and plans which may affect you or your job.

They will confirm details such as your job title and employee reference number to establish their credentials to speak on behalf of the company.

If you have any doubt about their authenticity end the conversation and call the office.

**Beware of the Media**
If the building is closed or damaged in some way it is quite possible that the media will want to obtain more information and some 'juicy' quotes.

Be careful when speaking to strangers who may try to make contact with you.

If they want information about the company, its activities or intentions you should refer them to the company's PR agency – Doolittle and Saynowt on 090 8794 5622.

**Rehearsals and practice**
There are regular fire drills which will be announced via the company intranet. Normally you will receive one week's notice and a daily reminder. This is to ensure that a drill is not confused with a real incident.

The business continuity team also run annual BCM exercises which may include a total or partial evacuation of the building. You will receive one month's notice of such and exercise together with the broad details of how the exercise might affect you. If your department is involved in such an exercise be prepared to take part in an evacuation on the day of the event. The final decision about evacuation will not be made until the exercise scenario has been revealed to the Emergency Response Team who may, or may not, decide to evacuate the building.

**PEEP Customising and Updates**
This is your own Personal Emergency Evacuation Plan which has to be tailored to suit your own needs in regard to evacuation of the building known as Beta Block, run by Elfin Safety.

Enter your details, where and as appropriate, bearing in mind the location, or locations, where you will normally expect to be based.

Individual floor plans can be downloaded from the 'BCM PEEP' folder located on the J drive where you will find complete instructions for tailoring this plan to your own precise needs.

Once the tailoring is complete you should print it as a double sided document and fold where indicated. Then it can be tucked away in the pocket of your overalls. You may wish to make more than one copy and keep them in alternate locations.

You PEEP should be reviewed and updated whenever there is a change to your working environment or duties. In any case it should be reviewed at least every six months.

**Useful Phone Numbers**

| Name | Number |
|---|---|
| Thomas Jones | 0123 456 789 |
| E Presley | 0123 456 987 |
| Jonathan Cash | 0123 546 897 |
| Dorothy Parton | 38 24 32 |
| | |

**Space for Personal Notes:**

**CAUTION**

PLEASE DO KEEP THIS PLAN HANDY AT ALL TIMES UNTIL YOU ARE FULLY CONVERSANT WITH ALL OF ITS DETAILS.
ITS CONTENTS MAY HELP YOU TO SAVE THE LIVES OF YOURSELF AND OTHERS.

I have also seen various folded versions of plans which open out from a small credit card size to a full page. These are usually printed up in bulk by specialists who have the necessary cutting and folding machinery. Normally, the cost and quantities involved in mass production prohibit the use of this format, except in those circumstances where there is likely to be a reasonably large number of people who will be in attendance on a short-term or temporary basis and may require a generic type of plan for the duration of their work assignment or course of study.

Where such people are likely to commence their attendance as a group, it makes sense to include the introduction and distribution of PEEPs during the orientation process. This would apply whether they are provided with a standard, preprinted generic plan or if they are expected to develop their own, in which case they should be provided with a word-processing template to get them started.

## A Practical Exercise: University of Lincoln

The University of Lincoln has to consider the needs of disabled staff, disabled students, and the possibility of disabled visitors, both known and unknown. The HR department is responsible for the policies and process with regards to the care of disabled staff. The university's HR department incorporates the assessed need for a PEEP through the staff probation and orientation checklist that is managed by line managers. This means that all new members of staff who require a PEEP can be identified and dealt with as an integral part of the induction process and at an early stage. Existing staff members who develop a disability can also make their needs known to their line manager, initiating the PEEP process. The student services department has a disability team with a robust PEEP process to take care of the evacuation needs of disabled students.

Fortunately, by embracing the requirements of disabled people within the university's BCP, it is possible to highlight their special requirements with senior management. In a recent exercise, the BC manager gave the incident management team an emergency scenario which identified a number of problems to be resolved. The university campus is a challenging site, divided by a railway line; as a result, the two halves of the campus have completely separate power supplies and different access routes.

The basic scenario was bad weather conditions causing transport and other difficulties. Over time, this led to a power outage in one part of the campus. Upon investigation, it was discovered that a disabled person was stranded on the first floor above the ground floor of the student services building. This individual was a wheelchair user. There is an external metal staircase exit from that floor as well as the internal stairs. Taking into account the weather conditions, it was considered a risk to try to make use of the emergency exit stairs. As this was NOT a fire, there was no urgency. It was deemed safer to operate the elevator mechanism manually. This exercise scenario had the full support of the disability team, a real disabled student, the health and safety manager, and the security team.

The strategy they adopted was to hand-winch the elevator up from the ground floor and then to reverse that maneuver to bring the disabled passenger down to ground level, thus enabling the person to get out to a place of safety. It took 45 minutes of hard manual work for the two members of the campus team to raise the elevator up one story. Although members of staff are routinely trained in hand-winching elevators across all of the campuses, the team had not participated in a full test to establish the time and resource parameters to complete a single evacuation using this process. Altogether, it was almost two hours before the disabled individual was able to get clear of the darkened building. These parameters can now be incorporated into future incident planning, employing simple mathematics to determine timelines for multi-story buildings.

**Lessons Learned:** Apart from the problem of the wheelchair above the ground floor, a number of other evacuation issues were raised during the exercise and the subsequent debrief. One of the factors which influenced their interest and concern was the large number of people who could potentially be involved in a full-scale evacuation. The university has up to 5,000 students who reside within close proximity of the Brayford Campus and a further 1,000 who live locally. Apart from the students, the university has around 1,100 support and academic staff, supplemented by visiting lecturers.

**Additional Evacuation Considerations:** One of the threats to the university site is the possibility of large-scale flooding which could be brought on by prolonged heavy rains. To cope with this sort of situation, the university has evolved a two-stage evacuation strategy in which the most vulnerable people leave the site whenever the flood risk is considered to be high and imminent and an evacuation instruction has been authorized and released by the incident management team. This first or pre-evacuation stage aims to get all of the disabled people, together with their care workers, off the campus and en route towards home or, if required, into temporary accommodation well away from the flood risk. The university has worked closely with both internal departments and external organizations to ensure that arrangements can be put in place to achieve a successful evacuation.

The university also recognizes the need for continuously monitoring the whereabouts and mobility of people throughout the campus. In other words, it is very difficult to know where everybody is at any one time. The movement of administration and service staff can be known with a fair degree of accuracy; most of them have a fairly static work style. Members of the academic staff tend to be rather more mobile and thus less predictable in their whereabouts. Students and members of the public form a volatile and mobile population which simply cannot be predicted or tracked. However, to offset these issues, the disability team works closely with each disabled student, reviewing timetabling requirements and assessing the unique needs of each student, as the student moves between planned academic events. All evacuation requirements for those students are diligently identified and exercised to ensure that they are effective.

One of the concerns of the university in this connection is the presence of a public right of way through the campus, making most of the buildings publicly accessible, apart from a few which incorporate access security measures. As a result, at any time, there could be an unknown, or undisclosed, disabled person somewhere on the campus.

Within a university, it is not feasible to employ a booking in and booking out procedure. Instead, the University of Lincoln has a skilled and trained team of fire wardens who are required as part of their role to ensure, to the best of their abilities, that all spaces within their area of responsibility have been cleared of people. This is then reported to the incident manager and forwarded on to the Fire Service appropriately. Discussions among colleagues continuously review the evacuation plans to incorporate the best methodologies to cover all evacuation scenarios.

## 4.5 High-Rise Buildings

The National Fire Protection Association (in the US) defines a "high-rise building" as one which is more than 82.25 feet (25 meters) high where the building height is measured from the lowest level of fire department vehicle access to the floor of the highest story which can be occupied.

**Background:** In 1852, Elisha Otis introduced the concept of the safety elevator, which incorporated a mechanism to prevent the cab from falling if the cable broke. The design of the original Otis safety elevator is somewhat similar to that which is still in use today. If the elevator should ever start to descend at an excessive speed, a mechanical device locks the elevator to its guides. Otis first demonstrated his clever new apparatus in 1854, at the New York exposition staged in London's Crystal Palace in what was described at the time as "a dramatic, death-defying presentation."

Three years later, on March 23, 1857, the first Otis passenger elevator was installed at 488 Broadway in New York City, although the first elevator shaft already existed in Peter Cooper's Cooper Union building, opened in 1853. Cooper included an elevator shaft in the design for Cooper Union because he was so confident that a safe passenger elevator would soon be invented. The shaft was cylindrical because Cooper believed it was the most efficient design. Otis went on to design a special elevator for this grand institute of higher learning. Today, the Otis Elevator Company is the world's largest manufacturer of vertical transport systems, i.e., escalators and elevators.

Without such devices, we would have very few high-rise buildings let alone huge skyscrapers. The competition to build higher and higher is largely fueled by elevator technology and innovative engineering. A major concern with all of these enormous structures is rendering them safe for the hundreds, sometimes thousands, of people who fill them on a regular basis. Such precautions involve preventative, protective, and reactive measures for all the various dangers which can be imagined. Lessons from the past provide valuable input to this type of thinking. Unfortunately, plenty of incidents in the 150-year history of tall buildings provide us with lots of evidence of what can go wrong and the likely costs and consequences of such events.

## 4.5.1 Categories of High-Rise buildings

From an emergency planning point of view, we can place high-rise buildings into two broad categories.

- Basic high-rise buildings – under 10 stories
- Skyscrapers – over 10 stories

**Basic high-rise buildings:** First, there are buildings which can be considered as a single entity where a single set of evacuation and other security measures can provide us with a comprehensive solution. The principal limitation here would be the number of stories involved. High-rise apartments of up to 10 stories existed in Roman times – long before the invention of elevators or escalators. There is no reason why we should not expect healthy people to make their way out of a 5- or 6-story building. It is quite common to find buildings of this size in which the only means of access to the upper floors is via a staircase. Presumably, it is unlikely that anyone with a physical disability would normally be situated above the ground floor. Therefore, you can rely on the stairs as a valid escape route in an emergency. Ideally, alternate routes should be provided, but many older apartment blocks have a single entrance and only one set of stairs. In these instances, it is obviously wise to invest in good fire detection and prevention measures such as smoke detectors and sprinkler systems.

**The higher the building, the more complex the problems become, with some of the super-tall skyscrapers requiring two or three hours to get everyone out of the building.**

**Skyscrapers:** Once we get to the second type of building, one with 10 stories or more, then you have to take into account that the majority of the people will have reached their destination through the use of electrical means, such as elevators, which would be probably be out of action in an emergency. Evacuation then becomes more complex because people may need to rest on the way down, and a proportion of the population is likely to be disabled with special emergency requirements of some sort. The provision of safe refuges, buddy systems, and various kinds of travel aids has to be considered in the planning process.

The higher the building, the more complex the problems become, with some of the super-tall skyscrapers requiring two or three hours to get everyone out of the building. Fireproof compartments need to be considered as a partial solution along with multiple exit routes, ideally one in each corner of the building leading to separate exits and different assembly areas, offering a choice of evacuation routes.

If you are considering the development of emergency evacuation plans and procedures from a skyscraper, you would be well advised to look at some of the reports on the collapse of the two World Trade Center buildings back in 2001. To compare, the World Trade Center towers were 1, 368 feet (417 meters) and 1, 362 feet (415 meters) tall. Out of the literally thousands of pages of reports, the essential parts of this material, from our perspective, have to be the recommendations for evaluation which come out of the research.

One of the key sources of information in this regard is the National Institute of Standards and Technology (NIST). In particular, I recommend the relevant sections of the National Construction Safety Team Act Report on the subject, NIST NCSTAR 1-7 and 1-8. Part 7 covers "Occupant Behavior, Egress, and Emergency Communication," while Part 8 covers "The Emergency Response." Both of these documents are included in the EEP Toolkit.

The full citation of the first these key documents is: Averill, J. D.; Mileti, D. S.; Peacock, R. D.; Kuligowski, E. D.; Groner, N.; Proulx, G.; Reneke, P. A. & Nelson, H. E. (2005), *Occupant behavior, egress, and emergency communication.(NIST NCSTAR 1-7)*, The National Institute of Standards and Technology.

The second one is: (Lawson, J. R. & Vettori, R. (2005).*Emergency response operations*. (*NIST NCSTAR 1-8*), The National Institute of Standards and Technology.

## 4.5.2 Evacuation and Escape Ideas

Emergency evacuation from high-rise, tall, and super-tall buildings is obviously a major concern for the designers, builders, owners, and occupants. The problem has been highlighted by the occurrence of numerous fires and other emergencies over the years, which have all caught the attention of the press. With these thoughts in mind, several people have come up with tentative solutions to the problem of getting large populations safely out of these towering structures. Some of these proposed solutions have been tried and tested while others have simply been spoken or written about.

**Tall buildings are here to stay, and we have to make them as safe as we can with the tools and techniques which are available to us.**

One school of thought considers that there is no absolute way of rendering such places safe for all of their occupants, especially in times when terrorism is so prolific around the world. This group suggests we should all stay away from jobs or tasks which require us to place ourselves at risk in buildings which are too tall to be safe. They also add the rider that such buildings tend to be iconic and are thus the preferred target of the terrorists who want to create maximum effect in the name of their cause.

However, tall buildings are here to stay, and we have to make them as safe as we can with the tools and techniques which are available to us. The principal problem seems to be that of vertical movement, i.e., getting down without the aid of the elevators or escalators which brought the people up in the first place. Horizontal movement across a level floor is not generally seen as a problem. Indeed, horizontal evacuation has often been proposed where there is sufficient space to provide separate compartments which can be regarded as fireproof long enough to enable people to access alternative routes out of the building. Of course, this is just a partial solution, but it is a technique which can buy time for those fleeing affected areas.

When considering horizontal evacuation as a part of the proposed evacuation strategy, it is important to call upon the services of qualified fire engineers who can provide professional advice and help to design and fit the structures and systems which might be required. Such experts will also have access to the tools which will enable them to gauge the timeframes for the various stages in such a strategy. These tools will also help to highlight any potential snags and delays associated with various configurations, enabling them to demonstrate which are the most effective and safest solutions. At the end of the day, however, they do not make the choice – they merely provide advice and information about good practice and its interpretation.

### 4.5.3 Escape Chutes

A number of companies offer a range of escape chutes in which the evacuee slides down a fabric tube or slide. Most manufacturers or suppliers offer them as a supplementary mechanism rather than the prime mover in an emergency. (They appear to have been developed from the laundry chute, which has been used for some time in high-rise hotels, apartment blocks, and hospitals as a way of moving dirty linen down to the laundry which is usually situated in the basement.)

The maximum flow rate of such devices is about 20 people a minute for users who are properly trained and organized; 10 people a minute for those who are not. If you wish to move a large number of people in this way, it may be necessary to install a number of these devices, a set for each floor involved.

Several variations on this theme are available. Sometimes they are offered as permanent built-in structures or, more commonly, as temporary devices which are to be deployed as and when required. One distinct disadvantage of these systems is the possibility of friction burns, especially where such devices are used to move people down several stories. To a large extent, this danger can be offset by the use of special clothing, such as jump suits, made available for everyone at, or near, their point of departure. This system is quite effective as a back-up measure in relatively low-rise buildings, but is not practical for use in the really high-rise situations.

I have also come across the suggestion of providing parachutes or even of using external pulley and hoist systems. Where there is an adjacent building of a similar height, it is sometimes possible to build horizontal evacuation bridges between the two buildings in order for people to transfer to the other building as part of the escape strategy. Clearly, this can happen only where the ownership and the architecture make it possible, but it has been implemented as a safe reciprocal strategy in instances where the neighboring buildings were part of the same complex and under common ownership.

## 4.5.4 Understanding EEP in the High-Rise Environment

The problems with evacuating a high-rise building mostly stem from the fact that one of the dimensions of the property is exaggerated. Therefore, the population density, the volumes of traffic, and the vertical distances demand a rather more intense and focused investigation, planning, and delivery cycle.

What can you do to assist or ensure the survival of people who might become trapped in the upper stories of a high-rise building? Long before the emergency planner appears on the scene it is almost certain that the design, engineering, and construction of the whole structure will have been finalized, and so it is too late to start thinking of working out alternative escape routes and exit points; they will be set in concrete, quite literally.

**If you do not develop and practice effective emergency procedures, it could be viewed as criminal negligence.**

Your starting point in this situation is to fully understand the structure and the thinking behind the layout and features which have been provided for easy and rapid egress for your population. Assuming that modern building regulations have been followed in the design and construction of the edifice, you will find that the escape routes and fire exits will be in place and suitable signage will indicate the way out of the building.

Ensure that your people can make the best use of these facilities and overcome the obstacles and errors which have so often contributed to confusion, injuries, and fatalities in the past. If you do not develop and practice effective emergency procedures, it could be viewed as criminal negligence.

So the question is: "What can you do to reduce the risk of an incident developing into a tragedy?" From a physical and mechanical perspective, you have to make sure that the right resources and facilities are in place. From an individual and group perspective, you must ensure that everyone is fully prepared, willing, and able to make the best use of what is available to them.

**Even if you are not faced with evacuating people from a skyscraper, this curriculum will give you useful information for planning the evacuation of employees and visitors from any multi-story building.**

## 4.5.5 A Practical Approach to Multi-Story Buildings

To address the specific issues associated with emergency evacuation in the high-rise scenario I have developed a curriculum which outlines a practical approach to implementing an EEP program within such a specialized and intense environment. It should be seen as a complement to the original EEP lifecycle, providing additional focus to the development and delivery phases of the program. Be advised that even if you are not faced with evacuating people from a skyscraper, this curriculum will give you useful information for planning the evacuation of employees and visitors from any multi-story building.

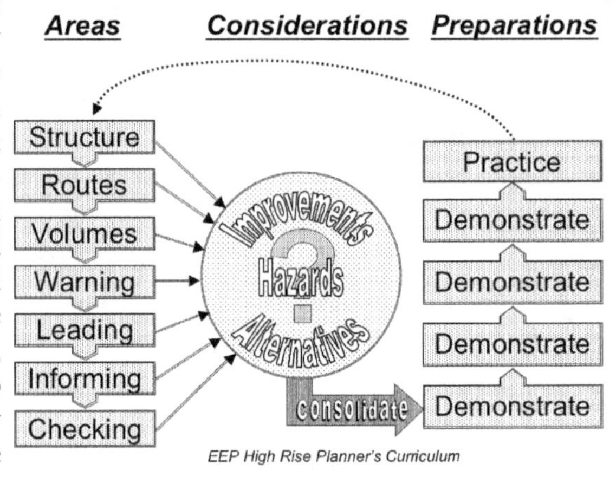


EEP High Rise Planner's Curriculum

In this curriculum, we identify seven areas in which you will need to spend more time and effort than you would for a regular evenly-dimensioned environment. Three groups of considerations need to be applied to each of those areas in sequence, followed by five stages of development for each. The end result will be a best-fit set of procedures and capabilities. Subsequent reiteration of this curriculum will ensure that the capability is maintained on an ongoing basis.

### *The Areas of Investigation and Concern*

1. **Structure.** Thoroughly understand the structured environment you have been presented with and become familiar with all of the details and their intended purpose in relation to emergency evacuation. This applies to every floor and section of the building.
2. **Routes.** Apply the same level of enquiry to gain an understanding and appreciation of the exit routes and exit points. Pay particular attention to the likelihood of them being compromised or obstructed. Once again check each floor and section of the building, avoid making any

assumptions, and think through the whole route from the point of view of an evacuee starting out from each of the locations you visit.

3. **Volumes.** Gain a clear understanding of the likely volumes of people who might be in various parts of the building. Visit and inspect each and every area, and remember to visualize growing volumes of people make their way downwards and mingle with the rest of the crowd.
4. **Warning.** Work out how everybody will receive, understand, and respond to the warning messages. Find out, or work out, what systems are in place or are needed. Don't forget to cover each and every area throughout each and every floor.
5. **Leading.** Consider how the growing crowds are going to be led, guided, or marshaled as they make their way through the confusion towards safety. Work out an overall strategy and develop detailed tactics for each group. This may be based upon floor plans, business groupings, or other common factors which particular groups might share. You have to know whom you are dealing with and how they might respond. Understand and workout who might be involved in doing the leading, guiding, or marshaling. They will need training.

**A very important aspect of evacuation planning... is that of accounting for every individual who was in the building at the start of the incident to make sure that no one is left behind, trapped, or injured.**

6. **Informing.** During the actual evacuation, keep the key players informed about progress and any further developments or special instructions. Figure out a suitable system and arrange for its implementation. Ideally, this should be a two-way communication system which will work under difficult circumstances throughout the whole of the building and its immediate surroundings.
7. **Checking.** A very important aspect of evacuation planning, especially in the high-rise scenario, is that of accounting for every individual who was in the building at the start of the incident. It may not be necessary to identify each and every individual but it is vital to make sure that no one is left behind, trapped, or injured.

## *Considerations*

As you carry out your investigations and begin to work out possible practical solutions, you need to consider three main aspects before moving on to the next stage.

1. **Improvements.** For each of the areas ask whether what is available or proposed is ideal or whether it could be improved in some way.
2. **Hazards.** In each of the areas keep a lookout for any potential hazards which might compromise the health and safety of the evacuees. A common hazard is the unauthorized use of emergency equipment and areas. The most prevalent type of abuse is parking trash or other unwanted or bulky items on the landings of the emergency stairwell. Apart from causing a dangerous obstruction, such items can also represent a fire hazard. Particular high-rise hazards include crowding and its consequences, dust, darkness, and smoke.
3. **Alternatives.** For each of the solutions which are planned or in place, consider all of the alternatives and whether they might offer an improvement which might be worth following up.

Repeat this thinking for each of the areas due to come under your consideration. Prepare a careful record of everything you discover, are told, or decide.

## *Consolidation*

Before moving on to the preparation and delivery stage, spend some time consolidating your thoughts and findings. Ensure that it all makes sense, the various elements are consistent, and there are no potential conflicts between the different interpretations of the strategy. If there are apparent conflicts, step back and review each of the steps, then eliminate the assumption, misunderstanding, or obstacle which led to the discord or inconsistency.

## *Preparation*

Once you have a clear understanding of the scope and scale of the problems and potential solutions, it is time to prepare for your population to find their way out of the building towards a place of safety in an emergency. To achieve this evacuation, people must know and understand what is to be done, they must have faith and confidence in the procedures, and they must have developed the capability to follow those procedures. You need to cultivate this knowledge, confidence, and capability both in individuals and in the entire community who share the same working environment. There are five steps towards this objective, which is to ensure the health and safety of the population whenever their building comes under threat.

1. Pull all of your ideas, considerations, and thoughts together and work out the overall strategy and the individual tactical details for each area or section of the building. Take the time to check that all of these tactical solutions can work either as

separate standalone procedures or as an integral part of the overall scheme.

2. Having worked out the solutions to the many problems, document your findings. This documentation will establish your intentions, act as a basic action plan for the next stage, and provide an audit trail. It will also prove to be a useful reference when you come to the next review and update phase.
3. Now that all the research and thinking has been done, it is time to put it all into practice. Make all the arrangements to allow the free flow of people from their point of origin through to the assembly area, including the development and delivery of all the support materials and resources, such as plans, escape aids, warning systems, and protective measures.
4. Once you are satisfied that everything is in place to enable a smooth evacuation, demonstrate the full procedure. This demonstration has two purposes. First, you have to prove that it does actually work in practice as opposed to the theoretical concept, which is what has been developed so far; this process includes checking the timeliness of the procedures. Second, you need to show people what you are expecting of them and thus convince them of the benefits and effectiveness of your EEP arrangements.

   This stage will probably require a number of demonstrations involving different areas and different groups of people. It is unlikely that a single demonstration will be able to reach the complete audience. It is best to start off with a relatively simple demonstration involving a small group of able individuals to prove that the principles work and present no obvious practical difficulties. From the feedback, you may need to make some minor changes to the way you organize subsequent demonstrations. Building upon this initial experience, you can then move onto working with larger and more diverse or demanding groups until you have shown everybody how it works.

   As we have pointed out in the regular or non-high-rise section, some disabled people may wish to be excused from participating in such demonstrations.
5. Once you have proved to yourself and others that the emergency evacuation procedures are effective, engage the entire population in a regular program of rehearsing or exercising their roles in an

emergency evacuation in the same way you would expect to conduct regular fire drills. The frequency of these exercises will depend to some extent upon the size and nature of the population. Consider the rate of change within the population or how long people stay in the same location.

Over time, you will need to get all of your emergency evacuation group to practice, communicate, marshal, work as a team, organize, and co-ordinate the evacuation procedures in order to develop the capability and certainty to be able to ensure the health and safety of all those who work, visit, or dwell within your high-rise building.

Having developed, demonstrated, and exercised your evacuation procedures, you cannot afford to sit back and rest on your laurels. The whole development and delivery process is subject to regular review and updating, as and where necessary. Don't assume that nothing changes. Continue to check for unexpected problems such as blocked passageways, cluttered stairwells, or locked doors. Make sure that the smoke detection, sprinkler, and alarm systems are kept in good working order.

It is also important that you keep up to date with what is happening in the way of new developments. Scan the horizon for any new ideas or technologies which you can employ to improve various aspects of the EEP. You should also make a point of noting any emerging hazards or risks which may lead to the need for refinements or alterations to your evacuation procedures.

## 4.6 Signs and Signage

No matter how carefully you plan, train, and test the emergency procedures, a large number of people will always rely upon the signs and notices to guide them to safety in the event of an emergency. It is possible that they may have read some of the small notices which are up on some of the walls, but it is unlikely they will have remembered exactly what those notices say. Under the semi-panic conditions of an emergency, they may be struggling to figure out which way is in and which way is out. The best thing under such circumstances is a simple, bold, and clear message.

### 4.6.1 Styles of Signs

While there appears to be no universally agreed international standard for signage in regard to safety, emergency evacuation, and related issues, a degree of commonality can be found in the style, shape, and coloring which is used around the world. Each of the signs has a broad purpose indicated by its shape and color and a specific message implied by a symbol representing the concept of the message.

## *Importance of Your "Unofficial Inspection" of a Site*

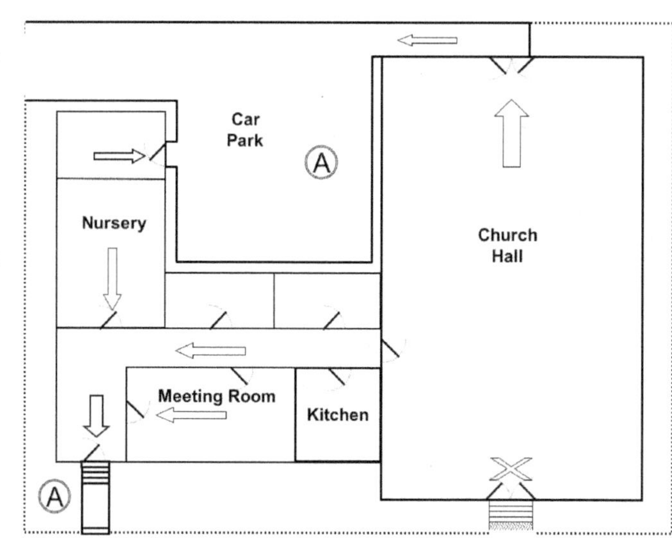


As part of my research, I have taken to reading all the emergency instructions in the various buildings which I visit on a regular basis. Prior to this, I have always assumed, as most people do, that I would always manage to find my way out to safety in the event of an incident occurring. I also adopted the common view that it wasn't going to happen when I was around. The term for this attitude is complacency, a widespread condition which often leads in the direction of negligence with regrettable consequences.

One of these unofficial inspections concerned a meeting room which my gardening club uses as a regular venue to hold talks and show prize specimens. This meeting room is one of the side rooms of a church, which also houses a children's day nursery and several other separate areas. The doors to the four entrances into different parts of the building are usually locked except when that particular area is in use. We enter and leave the building by the southwest door.

I found emergency instructions posted by each of the exit points and fire instructions alongside the fire extinguishers which were mounted nearby. A separate set of emergency instructions was posted on the bulletin board in the lobby area close to our southwest doorway.

Two separate emergency evacuation strategies seemed to be available. Only one escape route could be accessed from the main church hall, leading out to an assembly area in the parking area alongside the building. Apparently, the south door was not to be used in an emergency, presumably because it led straight out onto the road. The other escape route, the one for most of the rest of building, led out to a small assembly area on the grass between the building and the cast iron boundary fence. Actual maps showing the whereabouts of these two assembly areas were included only on the poster in the lobby; elsewhere, the assembly areas were simply described in words as either "parking lot" or "Ventnor Avenue."

Depending on which notice people might have read, those escaping from the nursery or our meeting room would either gather in the middle of the road (Ventnor Avenue), or they would congregate on the small patch of grass under a tree immediately outside the blazing building. Meanwhile, people escaping from the church hall would be standing alongside a blazing building surrounded by cars full of gasoline.

According to the instructions posted in the main hall and in our meeting room, "All members of the congregation who have a disability have their own personal emergency escape plans." I might be cynical, but somehow I doubt whether ALL of the disabled members of ALL congregations have a PEEP. However, such a notice might well be enough to satisfy anyone who is asked to check whether the church conforms to the relevant health and safety regulations.

The thought of a large crowd milling around in the dark among the parked cars while the church burns down around them reminds me of the scene I witnessed after a fire where several cars parked alongside a burning building caught fire before the firefighters had an opportunity to bring the blaze under control. At least one of the vehicles had exploded. If any people had been nearby, they would undoubtedly have been severely injured.

I have now arranged for our relatively small crowd to gather in a nearby neighbor's yard in the unlikely event of a fire on a Friday evening. In exchange, they now have a nice plant and are invited as guests to our December holiday party and our summer barbecue. One day they may even come around to our viewpoint and join the club. I raised my concerns with the churchwarden, who is going to bring the matter up before the next meeting of the Committee of Elders. The commander of the local fire and rescue service says the firefighters will close the road and move everybody well away from the scene immediately upon arrival. In fact, they plan to dispatch three vehicles: two to deal with the fire and one to deal with the occupants and the public. Hopefully the elders will modify their instructions to align them with the commander's intentions.

Many of the signs used in the church were the common off-the-shelf ready-made plastic ones with a small space in which to write a description of the assembly area with a ballpoint pen or a permanent marker. Such signs are common; they are an affordable and simple way of complying with the regulations and providing people with the information they might need. Unfortunately, the ink used in most of these pens or markers may fade and becomes illegible after a few years, especially where chemicals are used for cleaning purposes. One solution is to use something like plastic tape which lasts quite well or use enamel paint to do some sign writing. A third, short-term or temporary, measure would be to produce sticky labels using a laser-jet printer, but they are unlikely to withstand the attentions of an enthusiastic cleaner for very long.

The 5 broad categories of safety signs are shaped and colored like this:

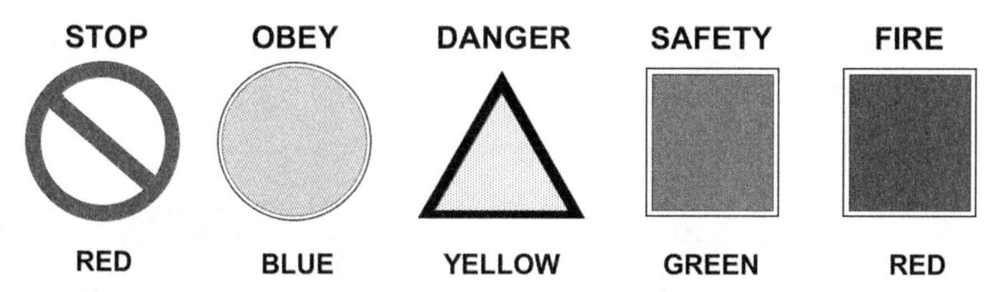


Generally five broad categories of signs relate to safety:

- *Prohibition* signs, which say "Stop" –
  a red circle with a diagonal red line and a black symbol.
- *Mandatory* signs, which say "Obey" –
  a blue circle with a white symbol.
- *Hazard* signs, which say "Danger" –
  a yellow triangle with a black border and a black symbol.
- *Safe Condition* signs, which indicate "Safety" –
  a green oblong or square with a white symbol or text.
- *Fire Equipment* signs, which indicate "Fire Protection" –
  a red oblong or square with a white symbol.

**Preference is given to symbols (also known as pictograms) rather than words because words may not be entirely clear to those with reading, learning, or language difficulties.**

It is the safety group which we are principally concerned with. Emergency evacuation signs are uniformly rectangular and green, bearing symbols or words which convey information about the route out towards safety or indicate the safe space once it has been reached. Green safety signs are also used to indicate the availability of first aid. Preference is given to symbols (also known as pictograms) rather than words because words may not be entirely clear to those with reading, learning, or language difficulties. Such difficulties will almost certainly be aggravated by the circumstances of a real emergency. Almost anyone can understand what a simple arrow indicates – the only question is, Which way is the arrow pointing?

Here is a typical emergency escape route sign, a plain green background with a large white arrow superimposed to show which direction the people should go. The white pictogram showing someone hurrying towards an open door implies that the indicated route will lead a person out to safety. The overall message of "hurry out in this direction" is adequately conveyed without the need for text. The color green means safety and the symbols suggest where it can be found.

## 4.6.2 World-Wide Developments

The original version of the "running man" symbol was designed in 1982 by Yukio Ota. It is now in common use in Japan and South Korea, referenced in the British Standard (BS 5499) and New York City local law (LL26) and is also used in Australia, Norway, and throughout China. However, it is worth noting that the actual implementations of this design do vary slightly. In North America, such exit signs are often white on a red background; local codes dictate whether the background color should be green or red.

In the UK, since 1996, text-only signs for escape routes and assembly areas are not permitted under the current fire regulations. All emergency evacuation signs must indicate their meaning through suitable symbols. These symbols are laid down in the British Standard (BS 5499). Within Europe, an EEC Directive (EEC/92/58) describes the symbols for emergency evacuation signs. The differences between these two standards are relatively minor; thus, there is little chance of confusion for the European traveler who visits or returns to the UK.

Australia has its own standard, "AS 1319 – 1994: Safety Signs for the Occupational Environment." In the US, OSHA publishes a fact sheet with concise guidelines about emergency exit routes and the associated signage. This fact sheet offers practical guidance on the implementation of the relevant part of the Code of Federal Regulations (part 1910-E of CFR #29).

Further details on standards and guidelines can be found in the chapter on Rules and Regulations and in the EEP Toolkit.

## Phase 4 Key Actions

- Conduct a thorough physical review of the site.
- Determine assembly area needs, select assembly areas, and arrange access.
- Develop and arrange escape routes.
- Understand and take account of how to meet the needs of disabled people.
- Develop and distribute personalized evacuation plans (PEEPs).
- Tailor plans for those with special, variable, and short-term requirements.
- For high-rise buildings, research the rules, regulations, and potential solutions.
- Make sure signs and signage adhere to local regulations, and specific site needs.
- Work out the answers to the "Hotel Pretend" scenario in the Discussion Questions.

# Discussion Questions - Phase 4

Phase 4 is the stage during which you determine the strategy on behalf of all those people who may be at risk whenever an emergency arises. Sometimes it is the point where we begin to realize the full implications and significance of the program upon which we have embarked.

The selected strategy, or strategies, will form the basis upon which all of the subsequent EEP plans, procedures, and protective measures will be built. So it is important to get this bit right. Perhaps it is worth stepping back for a moment and reflecting on what you have achieved and learned so far and what else needs to be done to achieve the desired end result. While I realize that you were asked to carry out a similar introspective exercise at the conclusion of the previous phase, I do feel that it is worth your while to spend a few moments seeking the answers to the following questions. If they give you any cause for doubt they may help to get you back on the right path. On the other hand, they may help to reinforce your confidence if you have managed to understand and interpret what we have been trying to convey to you so far.

## "Hotel Pretend"

For the purpose of this exercise imagine that you are preparing the EEPs for a large building in your vicinity, such as an office block or an apartment building. The important factor here is to be relatively familiar with the rough layout of the building and the environment within which it sits. This will make it easier for you to picture those elements which need to be taken into account when developing the EEP strategy. Let us suppose that this building is being converted into a hotel and you are responsible for developing and implementing the EEP strategy. From the size of the building, you can estimate the likely maximum population, taking account of guests, visitors, and staff. Of course, you could also think of a large hotel being converted into an office block or a residential care home. In any case, these are the kind of things you may encounter.

1. Work out the assembly area requirements for our "Hotel Pretend." Figure out the possible assembly areas that might meet your requirements and select those which are most appropriate. Select the associated escape paths, exits, and exit routes which you would recommend.
2. How many people will you need to plan for? Bearing in mind the variable nature of the population, what types of people do you think you should you plan for? How will this affect your proposed strategy?
3. Develop a strategy for the development and issue of personalized evacuation plans or PEEPs. Take full account of those members of staff whose work takes them to different parts of the building, such as maintenance personnel and house keepers. Consider the needs of long-term residents who may regard the hotel as their semi-permanent base.
4. How would you go about liaising with other interested parties such as HR, security, health and safety people, local authorities, and the emergency services? Figure out who these people might be and how you will make contact with them.
5. What signage will be required to suit the needs of the staff and guests? What signage might be required by local regulations? Draw up a plan showing which signs should be put up and where they should go.
6. Take full account of the characteristics and peculiarities of the location and its occupants. This will affect the types of strategy you should consider. Is it a high rise or large-scale building? Are the occupants likely to be transient, aged, or disabled? Take full account of any special needs which they may have.
7. What is the likely financial cost of your draft strategy for "Hotel Pretend?" What has to be done to meet the needs of the evacuees? Decide how you will persuade the management team to invest in the development and delivery of the plans and procedures. Prepare an itemized business case to support your argument.

# Phase 5

## Developing Plans and Procedures

Prior to the development of the EEP lifecycle model and its six progressive phases, I – and many others – would have regarded this area of activity as the starting point. Indeed, we would probably not have given much thought to the activities involved in the first four phases and plunged straight in at the deep end. As we hacked our way through the management jungle, we would doubtlessly have stumbled across such issues as developing an understanding and agreeing upon a strategy.

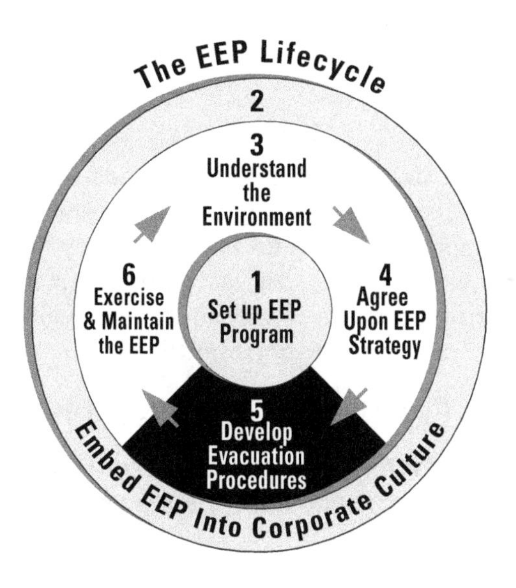


*This section will help you to*:

- ➢ Distribute evacuation plans effectively.
- ➢ Put in place procedures to ensure that everyone is safe and accounted for.
- ➢ Select and train the right people for your emergency response team.
- ➢ Anticipate and care for all the needs of those who have been evacuated.

With the benefit of hindsight, I can now offer a logical, phased approach. You can adopt that same approach without the need for a trial and error process. Through following our lifecycle model, by now you will have agreed upon an evacuation strategy which has been properly devised and constructed, based upon a sound policy and implemented in a professional manner. In the process, you will have begun to embed EEP firmly into the culture. With all this work behind you, it's time for you to think about the core process of developing and delivering the actual plans and procedures.

Prior to the EEP lifecycle model, I had already come to the conclusion that one should take a structured approach to the evacuation and assembly of large groups of people, especially because one has to take so many variables into account. These variables include the nature and size of the crowd, weather, environment, layout and location of the premises, type of incident, the scale and spread of the cause, secondary effects, the time of day, and so on. You can probably think of other factors which might impinge upon your plans and preparations. As you will see, the elements of my original structured approach all fit well within the more comprehensive and robust EEP lifecycle.

During this key phase of our work, we will be adopting a practical point of view, based upon real-life, hands-on experience. The suggestions presented are based on what works in the field rather than what "they" say one is supposed to do. This is not to imply that you should rebel against the advice of experts or executives. Rather, it is simply the need to focus on your main priority, which is the safety and welfare of the population for which you are assuming responsibility.

## 5.1 Evacuation and Assembly

> ***"Performance under pressure is a prized organizational skill. It will be remembered long after the crisis has passed."***
>
> Mark L. Friedman, *Every Day Crisis Management*, 2001

### 5.1.1 Preparation and Distribution of Emergency Evacuation Plans

*Emergency evacuation plans* (EEPs) are for use in evacuating the premises in the event of a fire or other emergency. This is called an *imminent catastrophic event* (ICE). ICEs are occasions for which there is some warning of a significant non-fire threat to a building or its occupants, such as a potential explosion, impact, collapse, or loss of access. Other trigger events might include flood warnings, power outages, industrial disputes, or crowds that are liable to become unruly.

Although evacuation procedures are often embedded within business continuity plans (BCPs), they are best understood as separate, standalone documents and should be used and exercised independently. The principal reason for this separation is that EEPs affect everybody who is likely to be in the building for any purpose, not just employees. Thus, the EEP applies to a much wider audience than the BCP and probably has a higher priority because its effectiveness has to be regarded as a key factor in the health and safety of all concerned.

**The EEP applies to a much wider audience than the BCP and probably has a higher priority because its effectiveness has to be regarded as a key factor in the health and safety of all concerned.**

Because EEPs and their procedures apply to a wide and varied audience, they must be relatively straightforward and easy to understand. They also need to be distributed widely so that, literally, everyone can become familiar with them and their purpose. The level of familiarity required can be achieved only by regular exercises in which everyone is involved. Those who are disabled or have special requirements may go through only an imaginary or virtual exercise, but they should certainly be included as they might be the most vulnerable in a real emergency.

Often, such plans take the form of posters and are on display throughout the building wherever there is an entry or exit point. (Entry points are sometimes known as access or ingress points, while exit points are also known as egress points). If these posters or signs are colorful and have a large or bold typeface, they will be more likely to attract attention and be read. Diagrams showing the routes and assembly areas will also help to catch the eye and convey a useful and memorable message.

In the United States, the main federal agency charged with the enforcement of safety and health legislation, the Occupational Safety and Health Administration (OSHA), has published a useful fact sheet "Evacuating High Rise Buildings" which gives some guidance to the occupiers and owners of high-rise buildings. OSHA defines a high-rise building as one which is at least 25 meters (or 82 feet) high. A copy of this fact sheet is provided in the EEP Toolkit. While this document is aimed at the US working environment, the information and advice it contains would be relevant to evacuating buildings of any size anywhere in the world.

## 5.1.2 Strategic and Tactical Planning – EMPs and ERPs

Much of your preparation and planning for an emergency should be carried out behind the scenes away from the view of the average occupants or visitors, who need to be aware only of the output from that process which strictly concerns them. Thus you will have:

- A strategic plan, generally referred to as an *emergency management plan* (EMP), which covers the whole gamut of emergency preparation and response thinking.
- These plans are supplemented by tactical level plans, generally referred to as *emergency response plans* (ERPs), which deal with the reactive phase of the incident, outlining the way in which the overall evacuation will be handled.
- Below these strategic and tactical documents are operational level plans which offer guidance and instructions to the end-user. These lower level plans may be issued in the form of posters and brochures for everyone to see and read; they may be distributed to individuals as an EEP; or a plan may be personalized for those with special needs in this regard, in which case it is known as a *personal emergency escape plan* (PEEP).

## 5.1.3 Operational Level Plans

Basically, your operational level plans are sets of instructions for people to follow in the event of an emergency situation which requires them to evacuate and go to a place of safety. As we stated in Phase 4 in the discussion of signage, you can display this type of plan as a set of simple clear instructions, posted where people can read and understand them quickly while on their way out. See Phase 4 of this book for the need for, and the content of, this type of generalized plan in our discussion of signs and signage, which covers the conventions for indicating direction, safety, and exclusions that make it possible to convey the essential messages without confusion or misunderstanding.

## 5.1.4 Adapting Plans

In all probability, you will take a template or a generic plan as your initial starting point when setting out to develop an EEP for your organization and the people who are likely to be on site at the time of an emergency. This will be much quicker and easier than starting from scratch, especially if this is your first attempt.

- ***Select a template.*** Discover a template or example which is best suited to your needs. For instance, if you are developing plans for a university campus, then it makes sense to work from an example which was developed for a university. On the other hand, if you are working with disabled people, then you will want to base your efforts on a plan which has been developed or adapted to meet the needs of persons with those particular challenges. Unless there is a compelling reason why you should use a particular plan as your reference model, it makes sense to acquire a number of samples and compare them. You are looking for a source document which is likely to suit the culture and style of your environment and those who occupy it. By comparing several, you may be able to use some of the ideas, layout and wordings from them to build up your own basic document. Otherwise, you will probably choose to work from the one that is closest to what you think your people would like to have.

- ***Decide on a format.*** If you are contemplating distributing the EEP as an online document, then you either need to select an on-line example as your model or you will have to convert your own ideas into a suitable on-line format. Perhaps a hardcopy document will provide you with enough clues for this to be a relatively simple task.
- ***Develop your tactics.*** At this point, you will already have an evacuation strategy in mind. The next step is to develop your tactics, which means exploring and surveying to select the exit routes, escape routes, and assembly areas and making enough notes to be able describe the evacuation procedure in sufficient detail to ensure that a stranger could follow the instructions.
- ***Enter the information in the template.*** Now that you have all the information, you will enter it into your template to form your draft EEP. (*If you have been working through Phases 1-4 in this book, you will have on hand much of the detailed information you will need for the various sections of the EEP.*)

## 5.1.5 A Standard Plan

For the simplest of situations you will be developing a single standard plan which everybody will use. This is where there is little or no variation in the type of people and they are all expected to follow the same route to safety. In this case the plan can be developed centrally and issued in its final form as a printed document, perhaps distributed by the HR or personnel department. Your major concern here is making sure that each person gets, or has access to, a copy of the EEP.

Some organizations have extra copies of their EEP or fire escape plans available in the reception area for visitors to read while they are waiting to meet their host. Be aware that this works only if the building layout is relatively simple and the evacuation procedures are straightforward. In such situations, the information can be conveyed on a small handout.

## 5.1.6 Generic Plans

When you need to consider more than one group, more than one class of people, or differing locations, you may choose to develop a suite of similar but modified plans for general use in different parts of the territory for which you are responsible.

To create a *suite of plans*, the process is much the same as the above except that it will be reiterative and the outcome will be general plans, similar in appearance but customized at the detail level to fit different groups, classes, or locations.

**Distribution of these plans needs to be organized to ensure that everybody concerned gets access to the right plans.**

Distribution of these plans needs to be organized to ensure that everybody concerned gets access to the right plans. The final version of each plan should be marked clearly to distinguish it from all of the others. Another possibility is to turn this type of plan into a sign or a poster which is displayed only in the location where it applies. Persons trying to escape from the right building with the wrong plan could be worse off than if they had no plan at all.

## 5.1.7 Tailored Plans

A series of tailored plans will be required wherever you have people with special needs. These are generally known as personal emergency evacuation plans (PEEPs) and we described them and their contents in the previous phase of our lifecycle. However, we did not disclose how these plans would be developed and distributed except to say that those who need them should be involved in the preparation of their own plans. (*For more about preparing and distributing PEEPs, see the Phase 4 section of this book.*)

Where such personalization is a requirement because of the varied nature of the environment, or the limitations of the people, then you will have to prepare a suitable template which can then be tailored to meet the needs of the user. The actual personalization can then be carried out by the user together with the person's helper, if he or she has one. To ensure that this process goes smoothly, you will need to provide some instructions about the personalization indicating what the person's options are. It is possible that some people with special needs may find it necessary to come back to you with additional questions or special requests. In any case, you should make yourself available to offer support and guidance as needed.

## 5.1.8 Emergency Evacuation Process and Timing

As you develop the plans and procedures for your EEP, you will need to develop a rational process for emergency evaluation and evacuation, which means you must make some assumptions or set certain parameters. This process involves a series of steps, which must be taken in quick succession with no allowances for hesitation. Based on experience and observation, we propose a model process bassed upon the following parameters:

1. There will be an *evacuation window* of up to 20 minutes.
   - This is the time from the first alarm to the incident occurrence.
2. The emergency assembly areas will be within a five-minute walk from the target building.
3. *Exit time* will be four minutes.
   - This is the time it takes to get everyone out of the target building.

While you cannot verify or influence the duration of the evacuation window, you can take steps to check the other two parameters. *If you are unable to meet either of these limits, then your organization must seriously question the wisdom of occupying that particular building.*

The emergency evacuation procedure might look something like this:

1. *An alarm message* is received and passed directly to the security officer, the person with overall responsibility for security and safety. This is not to be confused with the guard who does security – we mean the person with overall responsibility for security.

Target time = 2 minutes from ET Zero (Start of Emergency Time)

2. A *staff warning* is issued advising everybody that an emergency is being investigated. At this time, it is important to remind staff of your EEP policy regarding personal belongings – instruct them what they may or may not bring with them.
3. Meanwhile, the security officer will proceed to verify whether a genuine emergency exists. *Verification* may involve a call to the police or a quick review of the known facts and the current state of alert.

Target time = 4 minutes from ET Zero

4. Security officer then *confirms* the message and invokes the emergency evacuation procedure.

Target time = 5 minutes from ET Zero

5. Security officer *selects the assembly area and escape route* based on the information on hand. The default area should be indicated in the plan.

   Target time = 6 minutes from ET Zero

6. Evacuation is *announced* to members of staff by the most appropriate means. If there is no public address (PA) system, the message may need to be cascaded via floor marshals or wardens. The message should state clearly which exits to use, which routes to use, and which assembly area to use.

   Target time = 8 minutes from ET Zero

7. Everyone *leaves the building* via the selected exits, and security staff or floor marshals or fire wardens check that the building is clear before they leave. Everyone proceeds to the emergency assembly area using the selected route. Security staff and floor marshals follow immediately after the building is clear.

   Target time = 12 minutes from ET Zero

8. The whole group *gathers at the emergency assembly area* awaiting further instructions. Unless there are further developments, all heads of departments should conduct a headcount or identify anyone missing from those who were known to be in the building. All others should make every effort to assemble with members of their departments and make them known.

   Target time = 18 minutes from ET Zero

*Exceptions:*

- If the original alarm message is from a trusted official source, such as the police or fire department, *then the security officer should invoke the emergency evacuation procedure immediately (Step 5), as there is no need to verify that the alarm is genuine.*
- If the original alarm is not confirmed as a genuine emergency situation, then the security officer should issue a message so everyone can return to their normal duties with confidence.

This outline of the basic process would need to be tailored to suit any particular organization and its circumstances.

## 5.1.9 Emergency Evacuation Checklists

In the EEP Toolkit, you will find checklists that can be used to develop your emergency evacuation and assembly procedures.

# 5.2 Making Sure Everybody is Safe

> ***"I was involved in a real evacuation from a building recently because someone burnt some toast in a toaster...The smell of burning makes everyone think that there's a real fire, and they act as they would if the building really was on fire. So, hold an unannounced evacuation, and burn some toast to start off the event!"***
>
> Mel Gosling, independent consultant

## 5.2.1 Emergency Marshals

Throughout the whole of our emergency evacuation planning and thinking, the most important question which should be on our minds at all times is, *How do we make sure everyone is safe?*

You must make *absolutely* certain that *every single person* is accounted for and that all areas of the building are clear. It is for this reason that you will delegate the responsibility for ensuring all the people are evacuated successfully to specific individuals. Many terms are used to describe these key players in your health and safety arrangements. The responsibility may be assigned to members of the normal security staff as an integral part of their normal routine, or you may nominate other members of staff to drop their regular duties and adopt this role temporarily in the event of an emergency or an exercise.

These designated persons may be known as "emergency floor marshals" or simply "floor marshals," in which case they would be assigned to ensure that a particular floor or part of a floor is properly evacuated. Another term in common use when the focus is on escaping from a fire is that of "fire marshal," although the same techniques and principles would apply to any situation in which the premises need to be evacuated. We also come across the title of "fire warden" as an alternative.

According to the Merriam-Webster dictionary a *marshal* is "a high official ...usually in command of the military forces" or "a person who arranges and directs the ceremonial aspects of a gathering." On the other hand a *warden* is either "one having care or charge of something" or "an official charged with special supervisory duties or with the enforcement of specified laws or regulations; e.g., game *warden* or air raid *warden*."

From these two definitions, it seems at first glance that the title of *warden* is the more appropriate for the task we have in mind. However, a *marshal* would seem to attract more respect as a high-ranking individual who "arranges and directs." Perhaps the marshal is the one who would get the people out efficiently and quickly on the occasion, whereas the warden would make sure that they were looked after in the longer term, taking care of the administration as an integral part of the job. There is a school of thought which believes the term fire marshal applies to the person to whom everybody reports at the assembly area. This person is also the obvious source of information for, and the controller of, the assembled crowd. If all the other people involved in managing and controlling the evacuation are termed fire or floor wardens, then the fire marshal is regarded as the single point of control for the evacuation and subsequent movements of people. For example, it would be the fire marshal who announces the "All Clear" and tells people to return to work or to make their way home.

**What is really important is how these people carry out their duties rather than the label we put on them....The final choice of title has to be the one which suits the local culture.**

What is really important is how these people carry out their duties rather than the label we put on them. In some parts of world, the common titles are "fire warden" or "floor warden" and in other parts they are usually known as a "fire marshal" or a "floor marshal." The final choice of title has to be the one which suits the local culture; there is no absolute right or wrong in this matter. In the US, the Occupational Safety and Health Administration (OSHA) uses the term "floor marshal" in connection with evacuation planning.

In this book, we will use the term marshal to refer to the temporary officials with these responsibilities.

In practice, of course, we need to issue these marshals the distinctive tools they need for the job, which would include a high visibility vest or jacket. This obvious and distinctive piece of clothing serves two functions. Firstly, it makes them easy to spot and recognize, thus making it easier for them to gain everyone's attention. Secondly, it is a form of uniform and endows wearers with a certain amount of gravitas, making it easier for them to command a degree of automatic respect such that people follow their instructions rather than start to question the wisdom or the timing of what they are being asked to do. In most emergency situations, the majority of the people on site will have little or no idea of what is causing someone to interrupt the routine. It is less common for a disaster to strike in such a blatant manner that everyone is aware of the danger immediately.

Often the authorities, whoever they may be, are reluctant to reveal the whole truth to everyone about an impending threat because they don't want to risk panic that could aggravate the situation. Perhaps they don't even know the full truth themselves and are reluctant to reveal their ignorance. The belief that normal, sane people are likely to panic at the mere suggestion of danger is probably based upon imaginary fears derived from watching herd behavior among prey species such as antelope. Animals of this kind tend to flee from anything strange because that has proved to be a survival technique in response to the approach of a predator, making it safer to assume that anything unfamiliar has predatory potential. We humans, on the other hand, are not herd animals but hunter-gatherers, and as such we are usually inclined to behave in a more cautious and logical manner. In any case, we don't have the legs or the spring in our step to be able to outrun any serious predator; we have evolved to depend on our wits to keep us out of trouble.

Practical experience suggests that most people behave in a rational and calm manner when confronted by unexpected danger, providing they are in a relatively familiar environment such as the office, college, or supermarket. On occasion, however, people tend to act in accordance with the herd instinct. Herding behavior tends to occur where they have adopted a behavior to suit the place and the occasion, such as part of a football crowd. This herd behavior is exaggerated when the context encourages people to act as a mob rather than as individuals. They gather in groups, identified by their clothing and behavior, as fans of one particular team or another. We see the same effect in the audience at a pop concert, when large groups of people submerge their personal identity into the group identity. If something happens to suggest danger in such a situation, the immediate response is likely to be a fight or flight reaction because people in that situation are primed to think and behave as a herd rather than act as a number of separate individuals who think at different speeds with different views and experiences.

## 5.2.2 Two Emergency Evacuation Scenarios

We can identify two possible emergency evacuation scenarios. One of these you can handle as part of your normal business planning – the other will require that you be prepared to bring in specialized help.

- In the first scenario, when people are interrupted while engaged in the *normal routine of daily life*, we can expect them to try to analyze the situation based on the best information that is easily available to them and then react in what seems to them to be a reasonable manner to escape, avoid, or ignore the danger. When people are in this mode, we need to take steps to inform them in a clear and simple fashion and to provide them with clear and distinct instructions about where to go. We might call this the "inform and instruct" or "tell and guide" approach.

- In the second scenario, we are dealing with people acting in *crowd* mode, which means they have either gathered together as a vast group with a common purpose or interest, or they have somehow been forced together by circumstances beyond their control. Their behavior and thinking are likely to be at a primitive level, and they are likely to copy each other's behavior with the potential for a rush or sudden surge which could become catastrophic. When people are in this mode, they have to be controlled and directed in a much more organized or forceful manner. It is necessary to treat them like sheep because that is the way they are disposed to behave in these circumstances. Here we need a control and direction system which offers them no choices, no difficulties, and no obstacles in their way, thus ensuring that everybody automatically moves smoothly in the right direction. We might call this the "iron fist" or "no sympathy" approach. Generally speaking, evacuating such a crowd is a specialized task requiring the more martial tactics of an organized and trained police force. Dealing with large, frightened, or unruly crowds is way beyond the capability of the BC manager or emergency planner.

## 5.2.3 Pick the Right People

Effective emergency evacuation plans call for a number of marshals who will probably receive a minimum of training for the role which they may never have to play for real. Therefore, it is wisest to select the most suitable characters rather than the most convenient bodies. A really good way of selecting the best people is to run a series of exercises and to watch how different individuals cope. The trick is to do the selection effectively without giving offense to any of the parties involved.

**Taking part in such exercises gives potential candidates for the emergency response team an opportunity to self-select.**

I worked with one company that sent out a confidential questionnaire to all of the potential candidates for its emergency response team. The candidates were asked how they felt about taking on some of the responsibility for helping their colleagues deal with an emergency. The questionnaire included a brief description of the tasks involved and the sort of training they would receive. They were also told that they could retire from the training program after the first activity, which was to be a fire-fighting lesson in which they would learn about and practice with fire extinguishers. It was literally going to be a baptism by fire for the volunteers.

In practice, I have often found that taking part in such exercises gives potential candidates for the emergency response team an opportunity to self-select. During the subsequent de-briefing, I clearly recall Dorothy, who said that she felt rather overwhelmed by the situation we had just put her through, although the rest of the team had coped quite well with the evacuation of their eight-story building in response to a simulated bomb threat. A senior manager in an administrative department, Dorothy was respected as a very competent and capable person in her normal role. However, the frantic pace and chaotic atmosphere of uncertainty during the early phases of an emergency threw her completely off balance.

During the de-briefing she took a deep breath and said, "I felt most uncomfortable, especially at the beginning. My brain just doesn't seem to function at that speed and under that kind of pressure. I think I would be much more effective if I weren't given a front-line role, at least not during the early stages. Once the situation has stabilized and we know what we are trying to achieve, I am sure I could pull my weight, but I don't want to be a member of the A-team working under stress. I am sorry but that is just who and what I am, an administrator who thrives in a calm and orderly environment."

Another participant, Robert, took a rather different view, thriving on all that uncertainty. He was an IT support manager who was used to dealing with strange and out-of-line situations in which there were no simple or obvious answers. As he said of himself: "Over the years I have had to learn to think on my feet and I rather enjoyed the adrenalin buzz of the challenge which you threw at us. I like a bit of problem-solving; it's probably one of the reasons why I got into IT in the first place."

These two people, both highly competent managers, experienced totally different emotional responses to the unusual and stressful circumstances in which we placed them. Even though it was only an imaginary scenario, Dorothy experienced distress and found it difficult to cope with the uncertainty coupled with the pressure to make instantaneous decisions, especially where the health and safety of other people were involved. This is a well-known phenomenon, recognized by psychotherapists as a potential prelude to withdrawal behavior or depression. Robert, however, experienced what the psychotherapist would describe as *eustress*, a positive form of stress that is usually related to desirable or welcome events in a person's life.

The term eustress was first used by Dr. Hans Selye in 1975, when he published a model which divided stress into two major categories: eustress and distress. He described how persistent stress, or distress, may lead to anxiety, depression, or withdrawal. Symptoms occur when the victim is presented with what is perceived to be an overwhelming problem or set of problems. By contrast, if the stress involved happens to enhance the ability of the person to function by posing what is perceived as a challenge, it may be considered eustress. This condition has also been described as the "euphoria of stress," a process of exploring potential gains from a situation.

Both of these types of stress can be equally taxing on the body and are cumulative in nature. Dr. Selye observed from his research that the physical body has the same coping response to distress and eustress, the body responding to both types of stress via the hypothalmic-pituitary-adrenal (HPA) axis system. Therefore, it is important to give members of an emergency response team the chance to relax at regular intervals, not allowing them to remain under constant pressure for extended periods; fatigue and other symptoms will reduce their capability and may cause them to suffer uncomfortable after-effects.

In the case of Dorothy and Robert, we were able to adjust their emergency management roles to match their personalities without any one person losing face in the process. Robert became the leader of the incident response team, while Dorothy was given the role of administrator within the business recovery team to be activated after the trigger event was under control, operating at a rather more leisurely pace but with long-term objectives.

I have often found that people know themselves well enough to be able to self-select, providing we give them the opportunity to make the choice early enough in the training and development program. Do not force people into taking on responsibilities which are well beyond their comfort zone unless it is absolutely necessary. Generally, plenty of other candidates will be much better suited. In the long run it is wisest and safest to select your team members according to their capability and willingness, not in deference to their rank, title, or connections.

The key message here is that you should give all possible candidates the opportunity to explore their suitability and express their willingness in the relative comfort of an exercise or rehearsal. Take pains to avoid any suggestion of harsh criticism, thank them for their honesty, and make an effort to offer them roles in which they will feel reasonably comfortable.

On another occasion, I was facilitating an emergency evacuation exercise for the UK headquarters of a major international software company when my attention was drawn to some behavior worthy of note. This dynamic young company tended to employ young people who were deemed more likely to fit into its vibrant culture. It wanted to give its customers the impression that it was on the leading edge of technology and full of fresh new ideas. Consequently, many of the people involved in our emergency evacuation exercise were still in their late teens or early twenties. The dress code was business-casual from Monday to Thursday and relaxed-casual on Fridays.The identifying clothing for the fire marshals was bright red baseball caps emblazoned with "Fire Marshal" in large yellow letters. This kind of unconventional uniform suited the culture; the caps were convenient, fun for a young person to wear, and easily recognizable.

The exercise control team included a BBC reporter and a TV cameraman because we wanted to film the whole thing and use it as the basis of a training video so everyone in the company could benefit from the lessons learned. We parked ourselves in the foyer to the side of the main doorway in order to capture the evacuation scene without getting in the way. To make the scenario as realistic as possible, the exercise included actors dressed as firefighters and police rushing to the scene to put out the fire which was being simulated by the use of smoke generators. We were posing as the media who just happened to be on the scene at the time.

When the fire alarm went off, everybody sprang into action and people started pouring out of the building. We watched the receptionist don a fire-marshal's cap and start ushering people out of the building. However, although everybody knew there was going to be some kind of exercise that afternoon, we hadn't told them to expect a bunch of burly firefighters and a squad of police to arrive complete with wailing sirens and flashing emergency lights. As the receptionist approached us, she was obviously unsure about how to deal with our little group which was busily filming. She was torn between following her instructions, which called for her to clear the area, and dealing with the unexpected, which in this case appeared to be part of the exercise. As she approached us, she seemed to reach a decision to avoid us and followed the rest of the people out of the building.

However, that was not the end of the story. I watched her as she sought out the head marshal who was supervising the operation from the initial assembly area out in the courtyard. After she reported her area as "clear," she then explained about the strange group that was filming and asked him: "Is it all right for them to stay there, or should I try to get them to move away from the danger?" Immediately, she hurried back towards us and announced: "It's OK. You three can stay there."

**A combination of clear instructions, some thoughtful training, and choosing the right people had led to good teamwork in action.**

The point of this story is that this marshal followed her instructions as best she could, but she was uncertain when she came up against a bunch of strangers behaving in an unusual manner. Because her instructions didn't seem to allow for this kind of anomaly, she immediately completed her set task as best she could and then escalated her problem straight up to the team leader. Afterwards, she explained that she thought it would be quicker to check us out and come back with a clear strategy than to get into an argument with us. Apparently we looked pretty determined to stay where we were. The important point here is that she did all the right things: she kept a cool head, followed the procedure, and resolved the problem by referring it to a higher authority.

A combination of clear instructions, some thoughtful training, and choosing the right people had led to good teamwork in action. Consequently, they achieved a smooth, safe, and complete evacuation of the building, despite a minor diversion which was dealt with quickly and firmly. We came away feeling confident that this organization as a whole – and one individual in particular – was well prepared to deal with an emergency evacuation of the building. We had no reason to suspect that any of the other members of the team would have been less effective, since they were all products of the same process. All the people were out of the building and on their way towards a safe assembly area within four minutes of the fire alarm. It took about 25 seconds for the fire protection system to recognize the artificial smoke and raise the alarm.

## 5.2.4 Train Them

Because we can expect one or two individuals to withdraw from the program, and also because we need to have alternates who can stand in whenever there is a shortage due to absence or unavailability, it is wisest to recruit a surplus of volunteers who should go through the initial education and training program.

Once you have selected your initial starter group, you need to inform these volunteer marshals about the training program and provide them with a clear set of instructions regarding their new-found responsibilities and duties. Give them some time to get familiar with the instructions, and offer them the opportunity to ask any questions before moving on to the training stage. The training doesn't have to be lengthy or complex, but it does have to be sufficient to allow them to familiarize themselves with their various duties or tasks and to instill in them a degree of confidence.

Although I would not expect them to act with the skills of experienced firefighters in the event of a small office fire, I do advocate bringing in a professional firefighter to instruct everyone in the group in the use of a standard hand-held fire extinguisher, including some practical experience. Your local fire engineer will have the knowledge, techniques, and tools to be able to demonstrate the correct use of hand-held fire extinguishers. Such a demonstration should leave your fire marshals with a better understanding of fire and its effects, which will make it easier for them to carry out their part-time emergency duties.

The training for the volunteers should also familiarize them with the fire protection and alarm arrangements as well as any other means of emergency communication which might be available, such as a public address system or loudhailers/megaphones. It may also be necessary to train them in the use of all of these arrangements if it is considered likely that they may need to use them in an emergency. Remember, the more they know about what will go on during an emergency, the better they will feel about the task before them.

Once they are familiar with what they can expect and what is to be expected of them, you must allow them the opportunity to practice their new-found capabilities. Such practice will not only prove that the whole evacuation plan works, but it will also develop their confidence so they can rise to the challenge if ever they have to deal with a real emergency. To maintain that level of confidence, it is essential for them to have the opportunity to practice on a regular basis in an exercise, drill, or rehearsal. An emergency evacuation exercise should be conducted at least once a year, preferably twice a year. Apart from the benefit to those charged with various responsibilities, it is also a way of ensuring that everyone else is fully aware of what to expect in the event of an emergency. Crowds of people who know what to expect and where to go are likely to behave much more sensibly than those who are taken completely by surprise and could succumb to all sorts of fears. This is especially so if their knowledge is based on experience rather than having read an email or seen a notice posted on a wall somewhere. (*For more about practicing or exercising the EEP, see Phase 6 in this book.*)

## 5.2.5 Tools for the Job

Obviously, if we are to expect these volunteer marshals to carry out their duties effectively, we must provide them with the skills and knowledge. In addition they will need access to the tools for the job, which can be done in two basic ways: each floor or area might have a small cupboard or shelf where these items are kept and can be grabbed easily, or you may prefer to provide each individual with his or her own kit which will always be at hand at, or near, his or her desk or workstation.

If you choose to have a collection point where the marshal's kit is kept, then it can also be the site where the emergency exit plan for that area is displayed alongside a fire extinguisher and fire alarm call point.

Essential items include:

- A high visibility vest, preferable marked to indicate the role of the wearer.
- A flashlight, since in an emergency there may not be any lighting.

- A pair of gardening gloves to handle anything which is hot or sharp.
- A facemask to protect the wearer from smoke or acrid fumes.

Optional items:

- Loudhailer or megaphone, useful to get the attention of large numbers of people, to communicate in a noisy environment, or to direct people at a distance.
- Walkie-talkie radios, if they are already in use within the organization.
- Whistle to attract attention or summon help.

Whistles can be used to communicate a number of messages, providing you have a protocol in place with which everybody is familiar. You can use a combination of long, short, or multiple blasts to convey different simple meanings. For example, you could adopt an emergency code such as:

- One long blast = "I Need Help" or a Call for Assistance
- One short blast = "Take Care" or a warning of danger, such as a slippery surface
- Two long Blasts = "Pay Attention" or "Gather Round for an Important Message"
- Two short blasts = "This Way" or "Come Over Here," or "Follow Me"
- Three long blasts = "All Clear" or "Stand Down," or the end of exercise

Of course, the ultimate whistle signaling protocol would be to teach everyone Morse Code, which would permit them to transmit any message using their whistles, a rather old-fashioned and low-tech approach requiring considerable practice, which might not appeal to some of your colleagues. They would probably prefer to be issued walkie-talkies or mobile phones that might be of some practical use outside of the emergency situation. Walking around with a little tin whistle doesn't generate a lot of status, whereas a hard hat and a walkie-talkie do get attention and imply a certain amount of authority.

### 5.2.6 Making Sure No-One Is Left Behind

Amid all the confusion of an emergency it is always possible for someone to be left behind. Normally the fire and rescue service has the final responsibility to make sure that the premises have been completely evacuated. However, these professionals will naturally be reluctant to risk their lives by entering the presumably dangerous premises unnecessarily, and we should do everything we can to make sure that the building has been cleared of all personnel. Your intent should be to have everybody out before the professionals arrive on the scene.

More importantly, we should be able to prove or reassure ourselves that no-one has been left behind.

**Your intent should be to have everybody out before the professionals arrive on the scene.**

**Roll call and its limitations:** This assurance can be achieved by means of a roll call in which we check that everybody has reached the point of safety. One of the difficulties of this approach is that a roll call is a retrospective process which may occur too late to be able to find a person who appears to be missing. By the time we have completed the roll call, the building could have become an inferno.

**Knowing who was actually there:** Another difficulty is the uncertainty of who was actually in the building on this occasion at this time of day. Some large modern buildings have electronic security systems which count the people in and count the people out, but these systems can't tell us exactly where anyone is at a particular point in time.

| Total Continuity — Evacuation Checkpoint Card — Total Continuity | |
|---|---|
| **Fire Warden**<br>1. In the event of a fire alarm, or a fire drill:<br>2. Check that nobody is left in this area<br>3. Remove this card and take it with you to the Assembly area<br>4. Hand it to the Fire Marshal to indicate that this area is clear<br>5. It is important that nobody is left behind in the building<br>6. If you find anyone injured or disabled, shout for help or dial '999'<br>7. Do not leave anyone behind on their own | **Evacuation Area**<br>**First Floor; Gent's toilet**<br><br>**Evacuation Checkpoint Number**<br>**12** |

**Use of check-card systems:** Fire marshals are part of the answer because they can sweep their allocated area of the premises on their way out. They can also offer assistance to anyone who is having difficulty getting out of the building. Recently I was visiting our local recreation ground and saw something of interest in the sports pavilion. Alongside the normal fire instructions, which were posted on the wall in each room close to the door, a series of laminated postcard-sized cards were on display. Each one was numbered with instructions for the exiting fire marshal to scout the area, collect the check-card, and hand it in at the assembly area where the fire marshal would be able to note which areas had been cleared. Once the full complement of cards had been accounted for, then they could be quite sure that the building was clear of people. If one of the cards were missing, then the Fire and Rescue squad could be directed to the right area.

This evacuation check system was particularly well suited to a building like a sports complex in which several different and separate activity areas were being used by different individuals, teams and groups, many of whom were likely to be first time visitors. The whole site comprised two gymnasiums, four changing rooms, a shower block, an administration center, a basketball court, a running track, a multi-use sports field and a cricket pavilion. Keeping track of who was actually in or out of the group of buildings which makes up such a sports complex must be well nigh impossible. Not everybody signed the visitors' book when they went in, and even fewer bothered to sign out.

**Evacuation Area Checklist**

| Description of Area and Check Card Number | | Clear ✔? |
|---|---|---|
| Second Floor Board Room | 1 | |
| Second Floor Meeting Room | 2 | |
| Second Floor Kitchen Area | 3 | |
| Second Floor Open Area | 4 | |
| Second Floor Ladies Toilet | 5 | |
| Second Floor Gents Toilet | 6 | |
| First Floor Accounts | 7 | |
| First Floor Sales | 8 | |
| First Floor Administration | 9 | |
| First Floor Managers' Office | 10 | |
| First Floor Ladies Toilet | 11 | |
| First Floor Gents Toilet | 12 | |
| Ground Floor Reception | 13 | |
| First Floor Ladies Toilet | 14 | |
| First Floor Gents Toilet | 15 | |
| Ground Floor Stationery Stores | 16 | |
| Ground Floor Cleaner's Cabin | 17 | |
| Ground Floor Security | 18 | |
| Ground Floor Machine Room | 19 | |

These particular cards were attached to the wall using blue-tack reusable adhesive, but you could equally make use of self-adhesive hook and loop pads which can be obtained at most hardware stores or home centers. Indeed, that is the approach which I have now adopted in my own office and am recommending to my customers.

**Evacuation area checklists:** Presumably the fire marshal would be provided with an evacuation area checklist to note which areas had been accounted for and also to ensure there was a proper record confirming that the evacuation procedure had been successfully completed. I have drafted my own version of such a checklist which I use as a template or guide for those of my customers who are adopting this type of procedure.

In situations where evacuations or evacuation exercises are conducted on a frequent basis, a case may be made for the use of a board with hooks to hold the token which represents each one of the areas. It would be something like a key rack in a small hotel, which allows the receptionist to easily see which keys have been returned and which are still missing. Such a simple high-visibility tracking system can be especially useful at peak periods, such as during an exhibition or a convention, when large crowds of guests are coming or going.

A full page version of this checklist is included in the accompanying EEP Toolkit.

## 5.2.7 Refusing to Leave

A common problem for the emergency evacuation manager are those people who seem happy to ignore all the alarms and warning signals, usually on the assumption that it doesn't apply to them or to this occasion. They may also discourage others from leaving when they say something like, "Oh, that happens all the time. Let's ignore it and carry on. Someone will probably come and fetch us if it's serious."

Let us be very clear about this – such an attitude is downright dangerous. It could lead directly to someone being injured or worse. If it is a drill or a test, those who refuse to leave are falsifying the data which is being collected and thus misleading those who are responsible for security, health, and safety. In the long run, the resulting inaccurate data could lead to miscalculations or other errors in the planned response. Furthermore, people who do not cooperate are undermining the faith in the system which you are trying so hard to develop and maintain. Worst of all, they could end up putting their own lives in danger or the lives of those who are sent in to try and rescue them.

In the unlikely event that there is a fault in the system, then it must be fixed immediately, without delay – lives are at stake. If the alarm was set off by accident, then the whole population should be informed and precautions taken to prevent a re-occurrence.

Ignoring the fire alarm or any other safety communication should result swiftly in a severe reprimand from the highest level, and repeat offenses must be interpreted as punishable. This viewpoint should be expressed as an explicit warning in the terms or conditions of employment and should be enforced by managers and supervisors.

Failure to respond to the fire alarm has to be seen as totally unacceptable behavior except in those explicit circumstances in which certain individuals have been expressly advised to ignore the signals for a specific and valid reason such as, "an engineer is working on the system."

### 5.2.8 Tracking Systems

With modern technology, it is perfectly feasible to track everybody who enters an area such as a building, a business park, office complex, or a campus. Effective wireless systems have been developed for tracking goods in transit and tracing items of value such as laptops and other portable devices. GPS navigation is a common sophisticated application of such technology. While it may not be socially acceptable to force everyone who enters the site to wear a tag, you cannot rely on a system of optional choice to ensure that the premises have been completely vacated.

**While it may not be socially acceptable to force everyone who enters the site to wear a tag, you cannot rely on a system of optional choice to ensure that the premises have been completely vacated.**

When I discussed this issue with the security manager for one of my clients, she suggested two scenarios in which an electronic tracking system would be both practical and acceptable. She thought the prison service might use it to keep tabs on the activities and whereabouts of convicted prisoners. Her other idea was to offer this type of device to elderly or disabled people who might want to allow their caregivers to trace their movements in case of an accident. She was also reminded of a wildlife program in which radio collars had been successfully employed to keep track of a family of elephants wandering around a game park.

## 5.3 Evacuation Plan Content

### 5.3.1 Common Content

A few basic rules to emergency evacuation should be observed or reflected in all of your plans and planning.

In the event of a fire (or suspected fire):

- If you can see, smell, or suspect smoke, then you should crawl along the exit route low to the floor so you can breathe the cleaner cool air which will remain below the smoke layer.
- Before opening any doors, test them for heat by placing the back of your hand (not your palm or fingers) near, or against, the door surface. Do not open a hot door – find another exit route.
- Keep all “fire doors” closed in order to slow the spread of smoke and fire.
- Avoid using elevators when evacuating a burning building (unless the building has been specifically fitted with specially designated

fire lifts or elevators, which are designed, built, and commissioned for use during an emergency).

- Make for the designated assembly area and report to the fire marshal who will record your presence and safety.
- DO NOT return to, or re-enter, the building until the all-clear has been given by the authorities. If there is any doubt, remain where you are until given further instructions or information.

## 5.3.2 Strategic Evacuation Planning

One of the key elements of any evacuation plan has to be the assembly point. Preferably, you should set up a choice of assembly points to avoid problems with exclusion zones or accessibility. Assembly points should always be well away from the target premises since windows can be blown out from the higher levels of tall buildings and fall some distance away. In the City of London in 1993, I remember seeing shards of glass which had traveled about 300 meters (or about 984 feet) from the nearest damaged tall building. The Bishopsgate bomb, planted by the Provisional Irish Republican Army, caused the death of one person, injured 44 others, and caused £350 million (or over $572 million) damage. All together, we estimated that some 400 tons of glass had fallen into the street. It was a miracle that more lives were not lost.

Obviously, the taller the buildings in the neighborhood, the further the glass fragments are likely to fly; the location of your safe assembly points should take this factor into account.

You also have to remember that the emergency services vehicles will need clear access to the area surrounding the building and will require plenty of space in which to operate. In the event of casualties, then the medical and ambulance services will also need space in which to operate. It is not unusual for a designated convenient assembly area for evacuees to be commandeered by the emergency services and used as a triage or first aid area. Remember they take precedence in these matters. Once the fire alarm has gone off, the fire brigade has control over the "fire scene" until they hand it back.

As soon as the firefighters arrive, someone should be there ready to brief the senior fire officer who will need to know whether there are any hazards in the building and everybody is accounted for. The senior officer will not wish to risk the lives of the team unnecessarily in a strange and unfamiliar environment. The more information you can provide about the layout, the peculiarities, and any associated risks within the building, then the happier and safer everyone will be.

Unless obvious signs of immediate danger such as billowing clouds of smoke are present, people will tend to mingle immediately outside the exit door. This has the combined effect of slowing the evacuation and placing people within the immediate danger zone. Fire marshals should be appointed to keep people on the move towards the assembly area by reminding them of the direction to take and asking them to report immediately to the marshal who will be conducting a roll call to make sure everybody is safe and accounted for.

## 5.3.3 Triage

A number of emergency incidents come into the category of what I call "severe incidents," ones in which the damage is severe enough to cause harm to those present. Examples of severe incidents would include such events as explosions, storm damage, tsunamis, and earthquakes in which there is a chance of buildings collapsing or flying debris causing injuries and possibly deaths in extreme cases. Such events introduce the need for you to deal with multiple casualties or even mass fatalities.

*Triage* is the term used by the medics and the emergency services to describe the process of prioritizing the needs of multiple patients or victims. The term comes from the French verb "trier", meaning "to separate, sort, sift, or select." When dealing with the chaotic aftermath of a disaster emergency, medical efforts are further hampered by the sheer volume and variety of those calling for help. As people are being evacuated from the premises, their physical conditions needs to be evaluated. Those who do not require medical assistance must be moved out of the way quickly to enable medical personnel to deal with those who might be regarded as "casualties."

Basically, three categories of casualties can be dealt with by the appropriate level of trained helpers.

- Minor injuries: Those with minor injuries such as grazes and cuts are quite mobile and require basic first aid, after which they can join the others at the assembly point.
- "Walking wounded": While these casualties need more than basic first aid, the injuries are not life-threatening; they may require the services of a paramedic or a doctor but do not require surgery or transport.
- Serious injuries: These casualties are in need of rather more intensive medical attention, such as surgery. They will require surgery and physical assistance such as a stretcher or an ambulance to take them to hospital. Because of their special needs, these people cannot be moved from the scene of the incident before receiving immediate treatment. These more unfortunate victims become the focus of a rather busy area as

various services try to deal with different aspects of their treatment, care, and transport. In a really serious incident, you may encounter life-threatening injuries or even worse.

Dealing with this last category will take up a lot of space and will require a great deal of high-speed activity as emergency services personnel deal with each incident and its consequences. Their efforts will be severely hampered by the presence of anyone who does not need to be there. You need to plan to make the best use of space, which may be very limited, reserving areas for the different triage purposes, including those purposes which we hope will never become necessary. Unfortunately, history has demonstrated regularly that we do need to be prepared for the worst case scenario.

The emergency services in the City of London, for example, have a plan in place to enable them, at short notice, to erect a temporary mortuary which can accommodate up to 500 bodies. This plan has been tested in a multi-service exercise which set out to prove that they could cope with a large-scale severe incident. This plan takes into account that most of the available open space could be pressed into service as assembly areas for staff being evacuated from business premises in the area. On the other hand, the mortuary site is on private grounds which are of historic interest and thus is normally closed and inaccessible to the public; the emergency services would be able to gain access in collaboration with the owners of the land and the surrounding buildings.

## Geographic Triage at Hyatt Regency Disaster in Kansas City[1]

On 17 July 1981, in Kansas City, Missouri, in the multi-story atrium of the Hyatt Regency Hotel, the fourth floor walkway collapsed onto the second floor, and both walkways – crowded with people at the time – then fell to the lobby floor below, ultimately killing or injuring over 300 people amid the debris and dust. Dr. Joseph Waeckerle, former chief of Kansas City's emergency medical system (EMS), was asked to come and help out with triage at the scene of the event while the current chief dealt with what was happening at the nearby hospital.

When Dr. Waeckerle arrived, he found a group of more than 100 injured people on the circular drive in front of the hotel, the place like bedlam. Inside the hotel

[1] Friedman, M.L., MD (2002). Everyday Crisis Management: *How to Think Like an Emergency Physician.* Naperville, IL: First Decision Books. Anecdote derived with permission.

was even worse. According to the doctor, "It was like a war. There was a lot of screaming. Power lines had broken and were swinging above the lobby, arcing electricity. A waterline had ruptured, resulting in several inches of water on the floor. In it was floating feces and human body parts. You just had to ignore all of it and focus on what you were doing."

His first act was to perform what later came to be known as "geographic triage." By declaring that any of the injured who could walk should move to a designated area, he immediately sorted out the seriously wounded from those who were ambulatory.

Dr. Waeckerle grabbed a bullhorn from the fire department and took charge. Employing his new technique of "geographic triage," he instructed everyone who could walk (with the exception of the professional rescue teams) to get out of the hotel. "I had to explain to family members looking for loved ones that we could get to them faster if everyone went outside and allowed the rescuers to do their job...A number of people were buried alive. We had to bring in huge cranes to lift the debris in order to get to them." At times, the rescuers had to ignore hideously injured corpses, or even dismember them, in their efforts to extricate the living.

Adequate resources were not available to treat everyone at once. Altogether, 111 people died at the scene, 3 died later in the hospital, and more than 200 were injured, many of them seriously.

When asked to comment on particular lessons learned and the application of principles of crisis management, Dr. Waeckerle said: "People were looking for a leader, and it was relatively easy to assume command." Dr. Waeckerle's credibility was helped by his being well known in the fire and rescue community because of his previous position as EMS chief.

This technique of *geographic triage* in which the able-bodied people are asked to walk away from the scene towards a safer, quieter area for first aid and other purposes has also become known as "auto-triage." It has the advantage of removing surplus people from the immediate vicinity, allowing the emergency services better access to those who are in desperate need. At the same time, geographic triage helps to reduce the extent of trauma experienced by able-bodied persons being evacuated, since they are able to distance themselves from the more tragic aspects of the event.

## 5.3.4 Tactical Concerns

At the tactical level, a great deal of preparation and maintenance needs to occur before the event, and many actions which need to be carried out during and immediately after the emergency. Here are some thoughts to be kept in mind as you develop and implement the emergency evacuation strategy.

Volunteers should be posted at the assembly area to count people in and out of the premises. To make their task easier, they should be equipped with a clipboard. If possible, a list of staff members would be a very useful aid for them, but we do have to recognize that this is not practical in many cases.

As early as possible, you should make sure that someone takes the responsibility to assess whether additional help is needed and to make arrangements for that to happen. In most cases, you will have no need for additional support and this task can be ticked off on the checklist as "not applicable," but on those rare occasions when help is required, it is often a matter of "first come, first served" or "the early bird catches the worm."

Wherever there is a suspicion that the building or its facilities have been, or may become, damaged, someone should make sure that the gas, water, and electricity supplies are turned off. Usually this is one of the first things the fire and rescue people ask about, or they will simply go ahead and do it if they are familiar with the premises.

After the event, you can almost certainly expect an inquiry; perhaps a court case or an official inquiry at which evidence will be required. Therefore, it is vital to capture as much information as possible, including answers to the five standard media questions: *who, what, where, why*, and *when*. An official history log should be opened as soon as an *imminent catastrophic event* (ICE) is recognized or suspected. This log should be supplemented by as much supporting evidence as you can acquire. While such evidence would normally include photos, you may also include videos and sound recordings wherever possible. Note that it is almost certain that other people will be recording what they see and hear, and their material may be used against you, especially true now that hi-tech recording devices are built into the mobile phones that many people carry around as a matter of course. You should include at least one camera in the emergency kit and make sure that everything is photographed as soon as it safe to do so. You should also have some form of voice recording so that comments can be recorded at the same time.

Once the trigger event is over and everybody has been accounted for, the emergency services will be taking steps to render the place safe enough for personnel who are authorized and wearing suitable protective clothing to enter the premises in order to assess the damage in readiness for the restoration and recovery operations which may need to precede the return to work. In a major incident, this assessment may take up to 24 hours or even more. It is important to check with the emergency services that the incident is over and obtain specific permission from them before anyone other than authorized personnel enters the incident zone. You should also take account of their advice regarding appropriate protective clothing or any other precautions which they consider necessary.

After permission to enter has been granted, your reconnaissance team should enter in pairs and stay together while they are exploring and examining the building and its contents, recording what they see, hear, and smell as they go. A very useful item at this stage can be a long stick which can be used to probe while maintaining a safe distance from the object or area under investigation. At this stage, team members should try to identify and protect any evidence of deliberate damage or other unusual activities or items. I would recommend that one person should investigate while the other observes and records. They might take turns at either role, perhaps swapping over when they approach an area of specific interest to one or the other.

**Before you permit any of those who have been evacuated enter the premises, team members should arrange to make any damaged structures safe.**

Your early reconnaissance should also be used as an opportunity to check for any hazards and to identify the safest pathways or walk routes into and through the premises. If team members come across any leaking pipes, they should call for someone to cut off the water supply to that area providing it is safe to do so.

Before you permit any of those who have been evacuated to enter the premises, team members should arrange to make any damaged structures safe. Precautions might include highlighting danger zones, erecting safety barriers, placing hazard signs where necessary, and marking out any safe passageways. Sometimes moving furniture or other objects to a different position will help to make the environment safer either by preventing someone getting into a dangerous position, supporting a dangerous load, or simply maneuvering the object into a stable position.

## 5.4 Types of Premises

Premises come in all shapes, sizes, and complexities and, for much of the text, we have assumed that you are concerned with the health and safety of the people in a "typical" building, whatever that means. We have used a generality and avoided any attempt to specify what that might mean. In the previous section, we did look at how you apply the basic principles of EEP within a high-rise environment. The essential features of this situation were the large volumes of people and the complications introduced by the extreme height of such a structure. These considerations required a rather more deliberate reiterative process in which we gradually built a consolidated approach derived from a series of investigation projects each aimed at a specific area of concern.

I once worked with the facilities manager of a large industrial estate whose motto was: "Think big but work small." It was an approach which suited the size and mixture of his particular working environment. Obviously you have to tailor your approach to the peculiarities and endless variety of the premises and the population for which you are responsible. I can provide you only with guidelines and highlight some of the potential problems and solutions. As with the case of the high-rise scenario, the best way to set about things is to consider the various aspects and dimensions of discrete sections of the premises and the population as a series of standalone projects and then consolidate your conclusions and requirements as you build up the complete picture. In this context, I shall be referring to these target areas and populations as "evacuation areas" and "evacuation groups." A large or complex site might comprise a number of such areas, each with its own particular population.

Let us explore the various types of premises according to how their purpose, layout, and population may affect your planning. Finally, we must consider the special case of the secure environment which may bring additional challenges. For each of these situations, you need to adopt a structured, iterative way of thinking as previously described, but you should also be aware of some of the potential additional snags which you may encounter.

**Adopt the perspective of the endangered victims; try to see everything from the point of view of people who are experiencing the situation for the first time.**

When we talk about the types of premises in this context, we are concerned not only with the shape and structure but also its purpose or usage because that may influence the type and number of people and the way in which we communicate with them. Although every single building is unique, we can classify them into a number of categories, each with its own set of characteristics. This list is not meant to be exhaustive, but you should be able to align your area of responsibility with one of these categories. It is possible that you may see your building as a hybrid, or a poor fit, in which case there will be more than one set of hints or suggestions to which you can refer.

## 5.4.1 Large Building

Here we are referring to a building which covers an area so large that the population needs to be regarded as a number of groups, each of which will require its own version of the overall evacuation plans. The principal problem here is that of scale, and your principal assets will be patience and tenacity. You do need to be thorough and pay attention to detail throughout the planning process. Adopt the perspective of the endangered victims; try to see everything from the point of view of people who are experiencing the situation for the first time. They may be confused, frightened and disorientated. Their survival may depend entirely on your plans and preparations.

Divide the space up into discrete evacuation areas; develop tentative solutions; test and improve them individually; consolidate and co-ordinate them; test and improve them collectively; train and exercise EEP with the whole population. From then on it is a matter of regular reviews, exercises, and horizon scanning for any changes in the risk profile or the environment.

**One very important development will be...making sure that the multi-tenanted building is fully evacuated. Normally, tenants would be responsible for their own people...However, you may consider appointing someone to act on behalf of the landlord...to ensure the safe evacuation of the whole building.**

## 5.4.2 Multi-Tenanted

When use of the structure is shared, the situation is similar to that above, but the population is divided into autonomous and separate groupings. This situation presents you with the additional tasks of liaison, communication, and agreement among these disparate, possibly dissenting, neighboring groups of people. The onus will be upon you to bring these groups together and work out a plan of action which will result in a suite of EEPs which can be invoked separately or simultaneously.

The process will be similar to that outlined above with the additional initiation steps of needing to discover who your co-tenants are; establish contact with your opposite numbers; convince them of the need to work together for everybody's sake; agree how you will work together; and establish who is to take the lead. You must also consider whether the nature, style, or image of their business, or the way in which they operate, brings any additional risks that you might not otherwise be aware of.

During the development and delivery stage, you have to work out how the evacuation of the building is to be coordinated; you may need to consider prioritizing or sequencing the evacuation procedures (this would be determined by the special needs of some groups). You must also establish who will distribute the plans and how. The easiest approach would be to agree on a common template which can then be customized to suit each tenant's needs and culture.

One very important development will be the marshaling arrangements or making sure that the multi-tenanted building is fully evacuated. Normally, tenants would be responsible for their own people, a reflection of the legal perspective. However, you may consider appointing someone to act on behalf of the landlord as the overall fire marshal with the responsibility to ensure the safe evacuation of the whole building. If you should choose this option, do make sure that the arrangements and terms of reference are properly documented and you have a signed copy – just in case!

## 5.4.3 Factory

Factory premises can vary greatly in size and nature; so we can only generalize and must rely upon you to interpret our advice in a sensible and diligent manner. The majority of factory sites in most parts of the world will be subject to a wide variety of rules, regulations, and restrictions covering the way they operate, the materials they work with, and the goods they produce. Most factory sites will also include some form of warehousing to store supplies, raw materials, and finished products.

Particular problems which may give cause for extra concern in a factory environment include high noise levels which may make it difficult to draw everybody's attention to any form of alarm or alert systems. Also, additional health and safety problems can be caused by dangerous machinery, flammable or toxic substances, hot liquids or surfaces, and the movement of goods and materials. Such challenges usually result in someone being specifically appointed to take responsibility for these issues, and one would expect that person to be suitably trained and qualified for the post. One of your first moves will be to locate that person and develop a working relationship.

Locate the health and safety officer and, with his or her help, understand the particular risks and hazards; decide who is to lead the EEP program; agree upon the evacuation areas; prioritize those areas; investigate each in turn; determine suitable assembly areas; develop escape routes and evacuation procedures; develop, or confirm, emergency communications methods and protocol; document the resultant procedures for each evacuation group; test and review each procedure; conduct a full rehearsal or exercise; and set up an ongoing maintenance and exercise routine.

## 5.4.4 Business Park

A typical business park will accommodate a number of different independent businesses, most of whom will have little or no knowledge of what their neighbors do or how they do it. You might well become the first person to have knowingly made contact with all of your neighbors. Several years ago, I was employed by a disaster recovery service whose headquarters were in a small business park. Because of the nature of their business they tried to keep a discreet low profile. When I instigated a group emergency evacuation venture for the whole site, we were surprised to learn that two of our neighbors provided services which could benefit us – one was involved in logistics and helped us with a number of subsequent projects, while the other one provided technical support service which proved useful to a number of our customers. I understand that a number of other business relationships evolved from the sharing of knowledge which the project promoted. In addition, we all felt much safer and more secure as a result of our planning and the subsequent joint exercise.

Some business parks will be light industrial complexes while others will be more of the office type. Your EEP overall approach will be similar in either of these circumstances. Once again, the starting point will be the recognition of a number of discrete, separable evacuation areas and groups. Depending on the scale of the event and the size of the park, there may be an opportunity for you to develop some mutual shelter possibilities which will be particularly beneficial to any of the population who may be disabled or have reduced mobility. One of the difficulties might be finding sufficient assembly area capacity within the business park, since most of the space between buildings may be given over to busy access roads and parking areas. Preferably, a business park should include some open land to provide a security buffer between the occupied zones – but, of course, this ideal is rarely met and safe assembly areas can be difficult to find.

A sizable business park is likely to offer a number of entry and exit points, and evacuation of the whole community might be contemplated only as a reaction to extreme events. For example, if the business park is more than about half a mile across, it should be possible to adopt separate plans for various evacuation groups within the community. If you adopt this concept, then you must allow for incidents occurring in each of these evacuation areas. It is also likely that these groups and areas may overlap in such a manner that some or all of the occupants will have more than one evacuation option according to the nature, scale, and location of the trigger event. Multiple groups mean that your reiterative process is going to cover the same ground from more than one direction as you put your attention on various threat scenarios occurring in different places within or around your business park.

Start with locating whoever is responsible for security and safety within the park and establish what plans and preparations are already in place; contact the local police and the fire and rescue department for advice and guidance, since they will probably take command of the overall evacuation; arrange a meeting with your peers or someone who can represent each of the businesses on the park; decide who is to take the lead in this program to develop an overall strategy; investigate and agree upon suitable assembly areas, bearing in mind the capacity required; ensure that each business develops their own EEP tactics and plans; and arrange an exploratory desk-top exercise in which all of you play out a number of imaginary emergency responses. The aim of these exercises is to ensure that all the various plans and the options within them are compatible under all circumstances.

Agree to required amendments; develop an action plan; and arrange for a follow-up desktop exercise to confirm that the plans are now fit for purpose. Discuss your plans with the emergency services and ask their advice about the need for, and the wisdom of, an exercise or series of exercises to demonstrate the EEP capability to all concerned. Ask the emergency services if they will assist with, or take part in, your exercise program; it will help both parties to understand each other's needs and intentions.

You should also consider letting the local press know about your EEP intentions and endeavors; it will reinforce the importance of what you are doing and demonstrate your corporate citizenship. However, before going ahead you must clear this initiative with whoever is responsible for PR.

**...the predominant population in a shopping area will be temporary visitors who are not at all familiar with the place and certainly will not be available to take part in any of your training exercises.**

## 5.4.5 Retail Park

A retail park or a shopping mall will have many of the characteristics of the business park. The key difference is that the predominant population in a shopping area will be temporary visitors who are not at all familiar with the place and certainly will not be available to take part in any of your training exercises. They may also have children of all ages with them, and some of the visitors are likely to be disabled. If they are caught up in an emergency situation, they cannot be expected to know anything about your evacuation procedures. This unfamiliarity means they have to be given very simple but very precise instructions, and the signage has to be clear and easy to understand. Traffic will also be a major consideration when you are trying to figure out the escape routes. Assembly areas might pose another problem, and you may have to settle for a compromise solution and use areas within the parking areas as temporary refuges. Note that this will not be an acceptable solution if parking is in underground or multi-story garages.

I recommend that you engage the services of a qualified fire engineer and explore the evacuation possibilities using a good software tool which can provide you with a dynamic model of anticipated crowd flows and behavior in an emergency situation.

The actual process of developing a suite of evacuation procedures in a retail park or a shopping mall would otherwise follow more or less the same pattern as that described for the business park. Coverage in the local press could also provide some welcome publicity which would benefit all of the traders on the site.

## 5.4.6 School

In theory it should be relatively easy to plan for the evacuation of school premises. The pupils are all organized into small groups which are led by a responsible adult. Playgrounds offer temporary assembly areas, and the school register should enable the staff to know precisely who was in the building at the time. While a playground may serve as a temporary assembly area enabling an initial roll-call to be carried out, you will also need to establish a rather more distant option for those situations in which safety depends on getting the population well away from the danger zone.

Any program which seeks to address issues of health and safety of schoolchildren should receive unequivocal support from all those who are in some way concerned with the school. Your role will be to co-ordinate all that support and ensure a practical outcome. While the staff should have copies of the evacuation plan and notices will be posted in each classroom, I do not recommend that the individual children be provided with plans. It is far better for the children to be familiarized with the evacuation procedures through regular demonstrations and practice – this will lead to a much calmer response in the event of a real emergency situation developing.

The actual EEP process in this case is quite straightforward; the main concern in most school buildings will be the question of communication. It is important to be able to raise an alert without causing panic; sound the alarm; and keep the teaching staff informed of what is happening and expected. Drills and exercises can become a regular part of the school routine, ensuring that everyone understands what it is all about and that they are familiar with the process and the signage which will guide them to safety in the event of an emergency.

## 5.4.7 College Campus

A college or university campus is, in effect, an extended version of the school environment. The key differences from an average school will be a more mature and diverse population; larger numbers of people; multiple faculties, each catering to distinct and separate groups of students; and a wider geographic spread. There may also be a number of casual or visiting tutors who only attend for certain sessions.

Your advantage in this connection is the wide range of examples and templates which are freely available to act as reference material. In recent years, a large number of universities and colleges have developed their own EEP plans and procedures, and many of them have made these plans available to others. A number of examples are included (in the supplementary materials which you should refer to before setting out to develop your own campus plans). You may also find that the geography lends itself to a number of potential assembly areas, and it may not always be necessary to evacuate all of the buildings for every emergency; in fact, in some situations, alternate buildings might be used as sanctuaries or refuges. With so many variables, you will have to develop your own solutions based on a thorough knowledge of the site, the risks, and the population you are dealing with. (We did cover some aspects of this type of premises in *Phase 4* in the University of Lincoln example.)

The EEP process for such a university environment will follow along the general lines of that recommended for a business park, but it will have some similarities to the procedures for a school. In a college or university, it is likely that some of the students will be studying some aspects of health and safety, emergency planning, or risk management as a part of their curriculum. If so, then it makes sense to engage them in the process so that they may gain some practical experience and you have the advantage of several pairs of hands and eyes to help you with the investigation and delivery process.

From a project management perspective, you can consider two possible approaches for EEP in the campus situation. One approach is to appoint an EEP manager, representative, or administrator for each of the faculties and work together as a team under your leadership. The other way would be for you to act as the single point of control and center of competence. In the first instance, you will need to select and train your team. In the second case, you will need to collaborate with a number of people as you turn your attention to each particular area of the campus. To a large extent, the choice of approach will be determined by the culture, but the preferred approach will be for you to take the full responsibility for the development and delivery of the whole program. It is the route taken by the majority of such institutions.

The general approach which has worked well for others is to focus on one block or facility at a time and gradually expand your attention to include more and more of the campus until you have completed the first round of developing and delivering some basic plans and procedures. After that, it becomes a long-term reiterative process as you review, improve, and exercise. Each year there will be population changes with new students, faculty, and staff, who will require orientation into the EEP plans and procedures. PEEPs will need to be developed for some of the newcomers.

The prospect of a safe, well-organized environment should prove attractive to potential students, which should appeal to the administration that is always looking to enhance the image and reputation of the institution.

### 5.4.8 Theatre or Cinema

Under most legislatures, numerous rules and regulations seek to ensure the safety and wellbeing of members of the public who form the audience in theatres. Each of the levels of seating will have a number of alternate exit routes and exit points. However, once the audience members are out of the premises, they are either left to their own devices or they are under the control of the emergency services – at least in theory.

I would expect to see all of the exit routes well signposted and designed to ensure that the public can get in and out of the auditorium easily and smoothly. The phase of the evacuation in which I would anticipate considerable room for improvement is that of moving the people away from the building towards a safe refuge. Often, the fire exits of a theatre or a cinema lead straight out onto the pavement of a busy road.

Your challenge will be to identify a safe refuge within a reasonable walking distance of the theatre that evacuees can find quickly without getting lost. I know of a couple of such venues that have broad yellow lines painted on the pavement indicating the route to safety. When people are arriving, these lines also lead one towards the venue.

Information and instructions regarding the nature of the incident and evacuation procedures can be announced over the public address audio and visual systems which will already be in place.

### 5.4.9 Entertainment Complex

Here we are talking about places such as sports stadiums, fairgrounds, or festival sites. Generally speaking, the evacuation procedure is more or less the reverse of the entrance procedure, and the public address system will be the principal means of communication with the staff and the public. It is normal practice to have teams of ushers or attendants who are there to guide the public to their places on the way in, and they can also act as guides on the way out.

In the event of an emergency, one would normally expect the event to be postponed or cancelled; thus, the audience would be expected to make their own way home rather than seek refuge nearby. The attendants would be expected to check that their particular areas were vacated before leaving and then report to the head of security who would be liaising with the emergency services.

Communication between security staff and other officials would normally be through the use of handheld two-way radios. In the unlikely event that these are not available to the attendants, then you should insist that the organizers should invest in such equipment. It is not a major capital outlay, but it could save lives. Lack of such an investment would not go down well at the subsequent enquiry and prosecution.

**Ensure that people close all doors behind them as they make their way out...Open doors invite theft, fan flames, and allow smoke to add to the danger and confusion.**

## 5.4.10 Secure Environments

Secure environments make tempting targets, and an evacuation might be forced in order to gain unauthorized access to the secured materials or documents. Many motives could be behind such an attack (e.g., theft, commercial espionage, attacks of occupants, sabotage, protests, strikes, riots, or general unrest). Side effects could include theft of personal or commercial property. Your planning should take all of these concerns into account. Also consider the possibility that confidential information could be compromised, which could have serious implications.

If you are working in a secure environment, then it is essential that your evacuation plans and procedures must avoid introducing additional opportunities for breaches of security. From the outset, you must work in close cooperation with whoever is responsible for security to ensure that your ideas and actions increase rather than reduce the level of security during an emergency situation.

Wherever there is a clear-desk policy in operation, your plans and training must reinforce the need for all persons to ensure that all documentation or any valuable material is placed securely in locked drawers, filing cabinets, or safes before people move toward the exit. As an integral part of the exit procedure, all computers should be closed down, and any whiteboards should be wiped clean. Furthermore, you must take pains to ensure that people close all doors behind them as they make their way out. Emphasize that this is a fire protection as well as a safety measure. Open doors invite theft, fan flames, and allow smoke to add to the danger and confusion.

Another important factor in the secure environment is the need to account for every single individual without delay. Anyone who lingers or disappears could be using the emergency as an opportunity for mischief of some sort. I would expect a member of the security staff to be responsible for checking that all of the secure areas are vacated without delay. While security staff members will be ensuring that all the people are safely on their way out of the building, their primary concern would be security of the property and its contents. They will also need to check that during the confusion nobody slips into the wrong room by mistake or with disguised intention.

In a very secure environment, there might be sophisticated alarm or tracking systems to prevent people from by-passing security even in an emergency. Again, you do need to talk to your head of security to know what is possible, what is deemed likely, and what measures you need to take in order to maintain the appropriate level of security without compromising the welfare and safety of the population.

You must also make sure that the fire and rescue people understand the implications of any security measures which might appear to obstruct their response to a fire alarm or their ability to rescue someone. In the worst case scenario, emergency workers will smash their way into the property if they have cause to suspect that someone is in danger. On the other hand, if you can assure them that everyone is safe, then they will be content to stand back and simply manage the spread of fire or other dangers.

## 5.5 Helping People Afterwards

In the confusion and uncertainty caused by the sudden need to leave their familiar surroundings and rush off to gather in a strange place, wondering what is going on, it is quite likely that your people will have to confront various unexpected problems. The majority of these troubles will be a direct consequence of the incident or their reactions to it. Under normal conditions, most people will encounter occasional difficulties and learn to deal with them. However, the hurried circumstances of an emergency will tend to make them more error-prone and may weaken their resolve to sort things out for themselves. Also, they will be surrounded by others who have also made similar mistakes; as a result, you may find a strong tendency for an air of despondency to pervade the evacuated crowd. Of course, much of this can be avoided or reduced through planning and practicing, but with the best will in the world, you are unlikely to be able to prevent a number of unforeseen difficulties at the individual level.

## 5.5.1 Loss of Personal Property

During an emergency evacuation, many of those involved will lose track of some or all of their personal possessions. In the heat of the moment, they are likely to forget to take important items with them; their attention will be on survival rather than possessions. Since these bits and pieces of personal gear may not be close at hand, those evacuating may fear that retrieving these items may delay their exit, increasing the danger which they want to avoid. It is also very likely that some of their property may be damaged, destroyed, lost, or rendered unobtainable by the incident or its consequences.

The end result could be a whole series of problems that disoriented evacuees do not feel competent or obliged to resolve. They will be looking to their employer, or their host, to provide them with some assistance or guidance, and your tactical planning should take this factor into account.

Obviously, people who are injured, or taken ill, will be cared for by the emergency services, but in the process, they may be taken away from the site without their personal possessions or even their clothing. The priority for the rescuers will be to get the victims to a location in which they can receive medical attention as quickly as possible. Thus, the medical personnel will not be paying much attention to the personal possessions of someone who may not be coherent or conscious.

You need to ensure that someone makes a note of who among your population is taken away because of injury or illness. Let's call this person the rescue monitor. This volunteer should be provided with a clipboard and asked to jot down the details of everyone who is taken away or escorted to another location, creating a rescue record which should include details of who took each person, the reason why, and, if possible, the destination. Whoever is responsible for the welfare of those involved should make a point of following up with these "rescued persons" to make sure they have everything they need. Visitors and strangers should be included within this rescue monitoring procedure; you shouldn't ignore people just because you don't recognize them.

When preparing to deal with personal property problems, you might need to work out a policy regarding the handling of personal items in an evacuation and who within the organization is going to take responsibility for personal items. Consider how you might address such issues as the loss, or absence, of the following items.

- Handbags, wallets, or purses.
  - Apart from a whole bunch of items of sentimental value, the impact of a missing handbag, wallet, or purse could mean the loss of ready cash, credit cards, bus or rail passes, drivers licenses, passports, car keys, cell phones, address and phone number books, and front door keys. Each one of these losses could have serious implications for the person concerned, and without some form of independent proof of identity, the person may find it very difficult to obtain any substantial form of credit, support, or assistance from external sources.
  - The loss of a front door key might mean the person needs to employ a locksmith to force an entry into his or her own premises and replace the lock – not an easy thing to arrange at short notice, especially when you have no cash and no credit card. There could also be an issue with the police, who may suspect the motives of anyone who wants to break into private property without any proof of identity.
  - The loss of cash and travel passes or tickets could also pose a problem for those who are intending to use public transport to get home. I would suspect that walking home is probably not a viable option for the majority of your evacuees.
- Hats, coats, and jackets.
  - If people come away from their place of work or residence without their outer clothing, they may well have lost their wallets or purses and keys in the process, along with the protection and warmth of hats and coats. This exposure could pose a serious problem for them if the weather conditions are particularly poor.
- Cars.
  - In the aftermath of a serious incident, automobiles might be damaged, destroyed, or simply inaccessible. Without personal transportation, the drivers (and their passengers) are totally unprepared for the journey home. Even if they have their other possessions, they may not be suitably equipped or dressed to walk home in the wet or cold. It might be argued that they could take public transport, but probably they have driven to the location due to the lack of a suitable alternative.

- Diaries, calendars, and other personal documents.
  - Loss of personal records may make it difficult for the victims to adhere to social and business commitments in the days to come. While this may not loom as a huge problem, it could be worrying or embarrassing to the victims and may reflect badly on the organization which allowed such a thing to happen.
- Laptops and other valuable items.
  - Loss of valuables such as computers could be a very expensive inconvenience. Also a laptop that goes missing without being destroyed has security implications. If a laptop is saved but damaged, then recovery of the data is possible; there are companies that specialize in data recovery. Such services might be relatively expensive, but you have to take the value of data – personal and professional – into account when making such judgments.

## 5.5.2 Staff Help Desk

You should consider setting up an emergency staff help desk, or a support center, where people can come to look for help in such practical matters as those described in the previous section. Although makeshift arrangements are better than nothing, to be really effective, such assistance for staff should be arranged in advance, and everybody should be made aware. Perhaps the actual location of this function may be subject to a last minute decision based on the situation, but if everybody is aware of the intention, each person should be able find his or her way to it. Details of where the center has been set up could be posted at the marshaling point in the assembly area. Some pre-printed cards would be useful here, and they could be kept, together with some marker pens, in the box or cupboard in which the marshals' high visibility vests and other emergency items are kept.

For those who are faced with a temporary cash flow problem, you should have a float available to fund their journey home. You will have to make sure that recipients sign for the loan received so that the money can be traced. Repayment can then be made as a deduction from their wages. Hopefully, the organization would view these relatively small amounts as a sound investment in personnel welfare.

For those with a clothing problem, it may be necessary to set up some sort of purchase order to enable people to acquire outer clothing in a hurry. In the worst case, a few blankets might provide an expedient solution. I distinctly recall that when Hurricane Andrew struck the coast of Florida, one of our clients had to move a number of its IT staff at very short notice to a location in Canada where the weather was considerably colder. The company realized that tee-shirts were inadequate and provided the team with sweaters and warm jackets.

For those with a transport problem, you may possibly arrange car rentals at short notice or provide a taxi service of some sort. It would make sense if some sort of prior arrangements were made so that both parties understood the scale of the problem in terms of the number of people and distances to be covered. Terms of payment could also be settled in advance as it is unlikely that the evacuees would be able or willing to fund such temporary transport solutions. Another possibility is for someone to coordinate a driver and passenger scheme to make the best use of the cars that are actually available.

In the extreme case of a wide-area disaster, it is also possible that some people could be left homeless and will be seeking alternative accommodations. Your HR or personnel department should be ready to step in here with help and advice, liaising as needed with members of the local emergency services in the community.

## Managing Emotions in an Emergency

A disaster is a complicated event affecting numerous connections, intersections, links, and systems of people, places, things, and ideas. Disasters produce changes in human emotions that are both predictable and unpredictable. The regular ways of relating do not work during or immediately following a disaster. Normal cues are missing, images are distorted, and normal emotions and thoughts are temporarily incongruent. Mental health professionals who have been specially trained in disaster management know the unique needs of people in these distorted experiences.

Managers need to have at least a basic understanding of disaster exercises, tools, practices, procedures, and resources because they may be the only people available whom the staff trusts. If your staff does not trust you before a disaster, a disaster will not increase their sense of safety. If the manager is clueless and unprepared for a disaster, or has no concept about the effects of disaster on human emotions, chaos can increase and escalate the emotional consequences.

After a catastrophic event, it is often the quiet, centered, and calm voice during the event that a victim remembers. A calm voice of compassion that is resonating with reason and security is the loudest guiding force when madness is swirling noisily about, chaos is ripping apart the fabric of the known, and cacophony is jumbling up signals and signs that have before this moment made sense. Imagine for a moment the stairwells in the World Trade Center building during the earliest morning evacuations. People unsure of the situation quietly and quickly followed well-practiced procedures guided by managers reminding them of the drills they had done before.

Now, imagine a firefighter calling a child out from under his hiding place in a burning home. As the firefighter manages strong inner emotional content, the child hears a voice of authority, calm, and direction. Managing during a disaster is not about controlling the disaster; it is about managing the emotions of the moment. Disasters have a beginning, middle, and an end. Each stage is managed a bit differently. Importantly, there is a pre-disaster phase where the real planning is formulated and rehearsed.

*--Vali Hawkins Mitchell, Ph.D., LMHC*

Adapted from: Hawkins Mitchell, V. (2012). *The cost of emotions in the workplace.* Brookfield, CT: Rothstein Associates.

## 5.5.3 Reactions and Recovery

In an emergency requiring people to flee from an imminent threat, it is common for some people to become disoriented or even distressed in the face of the chaos. Certain additional complications can add to their stress. People may panic if they happen to get lost or find obstacles barring their way toward freedom and safety. Panic is made worse if people feel themselves to be abandoned on their own. Victims will adapt to these conditions more easily and tolerate them better if they are with others who can share the decision making and offer a helping hand when needed.

**Most people experience a sense of loss in this type of situation; not only have they lost personal items, but they also have lost the sense of security which they had until a few moments before.**

Once people reach a place of comparative safety and comfort, such as your assembly area, they will appreciate the opportunity to talk about what happened, what they saw, and how they felt. The need to talk about the experience is an instinctive reaction which helps people to defuse the tension that builds up under dangerous circumstances. It is a natural phenomenon and is to be encouraged. Their disorientation is made worse by the sense of mystery and confusion which arises when they don't know exactly what happened. Also be aware that most people experience a sense of loss in this type of situation; not only have they lost personal items, but they also have lost the sense of security which they had until a few moments before. Their normal place of work or residence is now associated with danger.

The most powerful way for human beings to deal with all these mixed emotions is to talk it through with somebody who can empathize with them. This is one of the key functions of a debriefing session which should be held as soon as convenient after some degree of normalcy has been restored. The explicit reason for such a meeting is to find out what occurred and whether any lessons are to be learned from the event. However, the side benefit for all those taking part is an opportunity to share their thoughts and emotions with other members of the group who went through the same or a similar experience. Your planning should allow for such a meeting to take place before anyone is expected to return to the scene which will doubtlessly remind them of what happened and trigger sad or morbid thoughts.

For most people, a debriefing among their fellow evacuees will enable them to deal with their emotional reactions to the event, and they will be able to return to the scene of the incident with the memory intact but inert. However, you may find some who will need rather more help to deal with their demons. Perhaps they have had previous similar experiences and it is the chain of forlorn memories which haunts them whenever they revisit what for them is now a fearful scene. These people will need the help of a trained counselor which is outside the remit of your EEP program. Your HR or personnel department should be in a position to offer help or advice in such cases.

## Phase 5 Key Actions

- Study the various types of plans to determine which ones suit your environment and population.
- Establish how long it will take to evacuate the premises for each of a number of scenarios.
- Decide how you are going to develop and deliver the various types of plans or instructions needed by your people.
- Develop or acquire templates to act as the basis for your plans, and distribute them to those who will be working on plan development.
- Take full account of the needs for triage and alternate tactics to allow for the unexpected turn of events.
- Select, train, and equip the emergency response team.
- Prepare and document methods for handling post-event problems, including loss of property and other needs.
- Work out how you are going to assist those who have been traumatised or injured.
- Adapt your approach to the type and size of the premises and the type of population.
- Develop and deliver written plans, involving people in developing their own personalized plans.
- Update your training and awareness program to reflect strategic developments and progress.

# Discussion Questions – Phase 5

Phase 5 will be considered by many to be the most important stage of the EEP program, although its success is completely dependent upon the efforts and outcomes of the previous phases. It can certainly be regarded as the most productive part of the program because this is the point at which the tangible and visible products that form the real benefits are generated and distributed. It is here that the subject becomes a reality for those whom we are setting out to protect. That means it is due to come under public scrutiny, and so it is wise to be cautious and check that there are no obvious errors or oversights which might invite criticism.

I would invite you to consider your answers to the following questions as a means of consolidating your knowledge and understanding. You must feel at liberty to challenge our ideas; we are not proposing that you should follow our guidelines blindly. Take them with a pinch of salt and adapt them to your own taste and circumstances.

1. How do you think you will approach the question of ensuring that everybody is safe and accounted for? Are there already arrangements in place which can be adopted or adapted to suit your needs? If so, are they tested regularly and is everybody familiar with them? How would you go about improving them?
2. Will you need to arrange for people to be appointed to various management and support roles to carry out the EEP procedures effectively? What will those roles be and how many will be required? Have you considered the need for alternatives or deputies? How will you go about selecting and training these people?
3. Are there any existing arrangements within your organization which might provide assistance, guidance, or counseling to those who have been evacuated? Will you need to develop or adapt some measures to deal with the personal consequences of an emergency, such as an unexpected evacuation of the premises? Do you think those measure will be adequate for all types and scales of emergency?
4. Have you thought about the evacuation arrangements which might be necessary for a large-scale incident such as a major flood or a storm? How will you liaise and co-operate with the emergency services in such an event? Do you know what their advice would be and how it will be disseminated? Is it likely that your people will all follow such instructions or guidance, or will they want to “do their own thing”?

5. Have you thought about who might support you in this important project? How will you ensure that you gain, and retain, their support? Will you need the support of more than one individual or department, such as someone to act as champion and someone to act as sponsor? Have you considered the need for a figurehead as well as the need for both financial and physical support?

# Phase 6

## The Ongoing Program: Exercise and Maintain the EEP

The ultimate phase of the lifecycle is the ongoing program. Most of the hard work has been done, but it will all have been for nothing if you do not ensure that the program remains alive and that everybody is involved in testing and exercising the plans to ensure they are effective and up to date. Embedding an idea or a concept into an organization's culture is an ongoing, continuous process. Ideas and concepts can be forgotten, ignored, or distorted easily if they are not refreshed regularly.

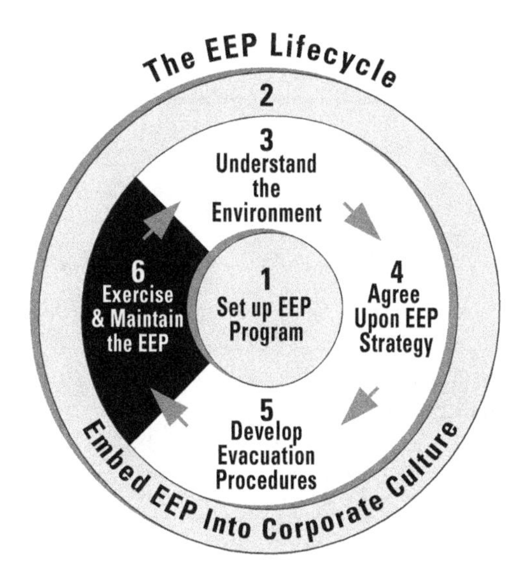

*This section will help you to:*

- ➢ Conduct a review that will result in an error-free plan.
- ➢ Differentiate between a "test" of your plan and an "exercise."
- ➢ Create an effective system to deliver, test, exercise, and maintain the plan.
- ➢ See the value of replacing conventional fire drills with "smart" fire drills.

Your principal concern in the ongoing program is the need to test and exercise the plans and procedures in order to be sure that everybody is familiar with them and that they are fit for purpose, including assurance that the resources and facilities are adequate. From a governance or audit perspective, your exercise and testing program should have an agreed upon schedule which would include regular reviews and updates of all of the plans and the resources required to support them. The need for such audits is especially true of the many personalized variations of the generic EEP which might exist. The frequency of actual exercises will, to large extent, be dependent upon the population which the program is designed to serve. For example, people with learning difficulties may need to go through some, or all, elements of the evacuation procedure on a frequent basis such as once a month; others may be able to maintain their knowledge on a quarterly basis. In any case, you should hold an evacuation exercise and a subsequent review at least once a year.

The ongoing program will also involve the development, posting and distribution of EEPs to all, and especially of the PEEP versions wherever they are deemed appropriate or necessary.

## 6.1 Peer Group Review

Before finalizing and distributing any of your draft evacuation plans they should be subject to a review process to ensure that they are:

- Accurate.
- Error free.
- Legible and readable.
- Fit for purpose.

*Accuracy:* The question of accuracy should, theoretically, be easy to answer. It is essentially a matter of going through the plans and, wherever there are statements of fact, such as directions, locations or names, they should be checked out by cross-referencing. In other words, check that what it says in the document agrees with what is recorded elsewhere or what is common knowledge. It is best to get someone else to carry out this type of review because of what I call "author's blindness." This is a strange mental block which can occur whenever you read something that you have written. As you read the text, you may subconsciously recall the writing process and see the original meaning which you were trying to capture at the time.

**An error-free plan is not misleading and conveys a correct message in the sense that the instructions lead to a correct conclusion.**

*Error-free:* An error-free plan is not misleading and conveys a correct message in the sense that the instructions lead to a correct conclusion. You are looking to eliminate or correct the mistakes that are rather more serious than a typing error which has been overlooked, for example, a plan which refers to "gather near a red light" when it should be saying "avoid all red lights." The best way to check for this level of mistake is for you to ask someone else to read through the plan and check that the instructions make sense and lead to a safe conclusion which is in line with the agreed upon strategy and tactics. This person may need to compare the plan with the physical layout, which may involve a trial run of the evacuation procedures.

Two kinds of trial run are possible. One type is what I would call a *cautious trial* in which the person carrying out the trial does so with caution and makes sure not to get into trouble wherever the plan is misleading or erroneous. The second type, the blind trial, on the other hand, would be one in which someone follows the instructions implicitly and unquestioningly with all the attendant risks. This latter type of trial is not to be recommended in connection with EEP.

***Legible or readable:*** Legibility or readability means simply that people can read the documents quickly and they will appear to make sense to them. In an emergency situation you have to allow for the fact that some of your audience may have reading difficulties caused or exacerbated by strenuous circumstances.

- In some cases, people will be using the plan as an *absolute guide*; you can assume they have no knowledge of the plan. They have to be able to pick up the meaning immediately without any doubts or mistakes. They must be able to trust that procedures will lead them to safety, even though they have no memory of having seen the plan before.
- On the other hand, if people are using a plan with which they are somewhat familiar as a *reference guide*, you can assume that they will recall what the plan is trying to tell them once they find the right section of the guide. In this case, users are familiar with the contents of the plan and refer to it only as a reminder to reassure them when they are doubtful about what to do or where to go. Such people have probably used the plan during an exercise or someone may have talked them through the procedures so that they have broad understanding of what the plan says, but they are unsure of the details.

In either case, you can only be sure about the legibility or total readability of a plan when it is seen through the eyes of potential users. Absolute certainty can be gained only when the plan is used under real emergency conditions and pressures, but common sense and a little imagination can give a reasonable level of reassurance. If you have serious doubts about an individual's ability to interpret his or her evacuation plan, then it seems that this person should be accompanied by a helper in the event of an emergency.

***Fit for purpose***: The last characteristic which the review process should be checking is whether the plan is fit for purpose. In other words: "Will it lead its reader to safety?" Here again, the only person who can answer this question with any degree of certainty is the likely user.

Taking all these considerations into account, it becomes pretty obvious that the only way to conduct a meaningful review of our plans is to conduct some form of peer review. Before finalizing and releasing your plans to members of the community in question, you should give them – or a representative selection of people from various sections of the community – the opportunity to review and comment upon the plans. When they are evaluating these documents, ask them to bear in mind the circumstances in which they are likely to be used. It might be helpful if you were to suggest the type of scenarios which they could consider.

The outcome from this review should be an affirmation that the plans are suitable for release or some recommendations for improvement. If major adjustments are required, you should consider repeating the review process. If it is only a case of a few minor improvements, repeating the whole review process may not be necessary. Once all the recommendations have been implemented successfully, you can go ahead and distribute plans to the whole of your population. This peer group review is a one-off task at the commencement of the sixth and final phase of the EEP program, a task which occurs before the full distribution of the plans.

## 6.1.1 Review and Update

Your strategies, tactics, and plans are liable to become outdated over time. They must be subject to regular reviews and updating so they do not become obsolete or misleading.

Theoretically, your thinking, arrangements and plans need to be revised only when there has been a significant change. In this context, a significant change is one which might impact any aspect of your emergency arrangements. *A significant change is one which will trigger necessary action within your review and update program.* Vague definitions tend to lead to oversights, misinterpretations, and avoidance; thus, you should adopt a default trigger which would act as a catchall, ensuring that you don't fall behind with your maintenance.

Where there is a regular audit procedure in place, it might be possible to include EEP currentness one of the areas which should be subject to audit.

Such audits might help to trigger your change management process, but the frequency of audit may not keep pace with the rate of change within the organization and its environment.

**Conducting a review of the EEP arrangements according to a timetable is the best way to be certain of capturing the implications of all possible types and sizes of change which might be considered to be significant.**

Conducting a review of the EEP arrangements according to a timetable is the best way to be certain of capturing the implications of all the possible types and sizes of change which might be considered to be significant. Thus, if it is a relatively stable establishment, schedule an automatic review on an annual basis. In a dynamic environment, increase the frequency of reviews to every six months. In either case, also allow for a review to be triggered by changes in working practices, the environment, the population, or rules and regulations.

The review process should cover the main stages of the EEP lifecycle. For example, check the policy, strategy, and tactics – are they still valid, or is there some reason why they should be modified to reflect current circumstances? If the answer is yes for the need for modification, then repeat that phase of the lifecycle and all subsequent phases. In those cases where these aspects are still applicable, then check whether the detailed information is still accurate such as names, contact numbers, and references to other source documents.

Follow up each and every review with a formal report complete with recommendations and an action plan to implement those recommendations. The formal report is important to provide a clear audit trail demonstrating due care and attention to the welfare of all those who are protected by the EEP program. If ever there should be a serious incident these records may be required as evidence in a court of law.

Whenever you update the plans or any related documents, you should make sure that you follow the change and control procedures in place in your organization. It is very important to make sure that all redundant materials are withdrawn from service and replaced with current versions.

## 6.2 Test and Exercise Defined

No plan of action has any value until it has been proven. Even then, it has precious little value until all of the actors have practiced their performance. There is no question that Shakespeare wrote good plays, but I can't imagine that any drama company would want to stage one of them without a few rehearsals. Remember, our actors are not accomplished professionals and perhaps we should not put too much faith in a plot that has not yet stood the test of time.

Seriously, we must test our plans to see how well they work. We must also challenge the assumptions about timings. Once we are reasonably confident that the plans should work, we must carry out a dress rehearsal to make sure everyone knows what to do and how to do it. Over time people will either have forgotten, lost confidence, lost interest, or been replaced. Thus, we should carry out practice evacuations on a fairly regular basis; otherwise, our plans could cause chaos rather than save lives, the latter being, after all, the whole point of the program.

You cannot have any real confidence in your plans and procedures until they have been fully tested. A *test* will verify their usefulness. While a test is an important step, it still doesn't prove that everybody will reach safety successfully. *Exercises* are the only way we can be sure that the people will be able to interpret the plans and procedures correctly within the requisite timeframes under difficult circumstances.

**A *test* establishes, or measures, facts and figures; an *exercise* develops or demonstrates atitudes and understanding.**

I'd like to take a moment to clarify the terms being used in this section; otherwise the whole idea of whether a certain set of activities is "testing" or "exercising" your EEP can become hopelessly confused. In older literature in the field, you will see that 20 years ago, any efforts to determine if a plan would work under real-world conditions were referred to as "tests" of the plan; however, the whole industry has "grown up" since then, has become more sophisticated and, along with it, our use of language. Current terminology in emergency management, disaster recovery, and business continuity includes these terms:

> **Test:** A *test* is a procedure which determines whether something works or is fit for purpose, and it produces a specific answer. Successive testing does not lead to improvement; like an audit, it merely reflects the current status. For example, testing can determine if certain aspects of the plan are still accurate and if anything has changed, such as names of people and companies, contact phone numbers, building layout, vendors of services, etc. These are all aspects of a plan that need to be tested for accuracy before any kind of real-life exercise takes place.

For example "I weigh myself to find out if I've lost weight, but I exercise to make the change happen."

**Exercise:** An *exercise* is a procedure, routine, or drill which is carried out for training, learning, and improvement. Capability should improve through exercising. Capability tends to diminish without exercise; regular exercising helps to sustain capability. A full-blown fire drill is an exercise in which, under the most realistic conditions possible, people act upon the instructions in the EEP to practice carrying out the tasks and routines that would be expected of them in a real emergency. The hoped-for result of an exercise is overall improvement.

For example "I exercise to keep fit and lose weight. It is an ongoing process."

Thus, a *test* establishes, or measures, facts and figures; an *exercise* develops or demonstrates aptitudes and understanding.

Start building your confidence by carrying out a series of tests. The aim of these tests is to prove that the procedures, facilities, and resources are suitable to support a successful evacuation under emergency conditions. If a procedure, facility, or resource proves to be inadequate or unsuitable, then something needs to be changed. It might be considered to be a failure, but it is an impersonal object or set of instructions which has failed. No person is at fault.

A number of elements or steps within the typical evacuation procedure need to be checked out and demonstrated, and we can use the same sequence for both our testing and our exercise program.

1. **Familiarity** with the purpose and any reference points.
2. **Invocation protocol,** which embraces the alarm and communications process.
3. **Exit,** which includes the choice of exit and exit routes.
4. **Escape routes**, or how to get away.
5. **Assembly area routine** which includes the ongoing communications procedure about what comes next.
6. **Stand down and return procedure** together with any caveats which might apply.

## 6.2.1 Element Testing

Each of the key elements within the evacuation procedure can, and should, be checked out separately before embarking on a full scale end-to-end test. In fact, if all of the elements follow a logical sequence then it may not actually be necessary to carry out a complete test; you could move straight on to the exercise program which will prove that the complete procedure is effective.

- *Awareness*: The first element is about making sure that all participants are fully aware of the purpose of the emergency evacuation procedures and their point of contact or reference source if they should require further information or assistance. This awareness can be tested by checking with a small group and asking them if they understand why the EEP exists and if they can figure out whom to ask or where to go for further information.
- *Alarms and communications*: Testing the alarm and communications procedures is relatively straightforward. Make sure that every person can hear or understand the alarm and the associated emergency communications system. It is very important that you give plenty of notice to everybody that there is going to be a test and a reminder nearer the time so that you don't cause an unnecessary panic. When you are announcing the test make sure you tell them how to react. If the test is combined with a fire drill, you may want everyone to leave the building, in which case the test is a success if everyone does leave the building. This proves that they heard and understood the alarm. Alternatively, you may ask them to remain at their post; in this type of test, you will have to question them about the effectiveness of the alarm and communication procedure. You can question them through an email survey, speaking to a small sample, or inquiring via their managers or supervisors. The choice of method will depend to a large extent upon the size, nature, and culture of the organization.
- *Exit routes*: Exit route testing is a simple matter of following the instructions given in the plans to check out the route from the typical starting points to the exit and ensuring that a typical member of the resident population is able to find and use the exit. This procedure should include reference to all the signs which should be there to indicate the route to the nearest fire exit. Check that the signs are easily visible, correctly located, and make sense to someone who is not familiar with the building. Where there are alternate routes through the building or alternate exits, also verify the selection and announcement procedure or protocol to make sure that you don't end up with conflicting flows of people following different routes or moving in the wrong direction.

- *Escape routes*: Escape routes, which are the pathways leading from the exit point to the safe assembly area, should be tested in a similar manner to the exit route; they are the external continuation of the exit route. When testing the escape route, you should bear in mind the volume and type of traffic which you can expect in an emergency and consider the possibility of all sorts of weather conditions. If you are expecting people to be in wheelchairs, then make sure that they can cope with the surface, any slopes or inclines, and avoid any steps or stairs. If steps cannot be avoided, then robust arrangements should have been made for them to be carried up or down at these points on the route.

**Your responsibility for EEP does not end when all the people are out of the building. It ends only when all the people are safely back in their offices, have reached their homes, or have been handed over to the care of some other agency.**

- *Assembly areas:* Assembly areas need to be checked out for availability at all times and measured for capacity to easily accommodate your maximum crowd size, including an allowance for visitors or guests. If the capacity and availability are sufficient, check out any special features you require, such as a marshaling point. Finally, make sure that suitable signs are in place to guide people to the correct area and provide them with any information which may be pertinent.
- *Return process*: The final element which you might want to test is the withdrawal or return process. How is the stand down going to be announced to the gathered crowd? If the people are to be sent home or moved to another location, who is going to inform them and how? This part of the evacuation process is often overlooked, ignored, or left to an ad-hoc arrangement to be determined when all the fuss has died down. However, such arrangements cannot be left to chance. ***Your responsibility for EEP does not end when all the people are out of the building. It ends only when all the people are safely back in their offices, have reached their homes, or have been handed over to the care of some other agency.***

## 6.2.2 End-to-End Testing

This is the ultimate form of testing and implies following the complete evacuation procedure including all of the elements mentioned above. I would only consider embarking on this scale of test in the event that the evacuation procedures appear relatively complex or the resident population is a particularly vulnerable one. In most instances, I recommend that your first full-length *exercises* be regarded as your final *end-to-end test*. Once, you have proved that all of the individual elements of the evacuation process are valid by conducting the series of recommended tests detailed above, the only remaining question is whether people can follow the instructions and complete the full evacuation procedure without exposing themselves to danger or getting lost.

**In most instances, I recommend that your first full-length *exercises* be regarded a your final *end-to-end* test.**

## 6.2.3 Exercising

Exercising in this context has two main purposes:

- A demonstration which confirms that the procedures do work.
- An education which teaches people what to do and how to do it.

Success in following procedures will develop their competence, which will, in turn, give the people confidence that they will be able to reach safety if it should ever be necessary. Confidence and competence will thus ensure their survival capability.

Because an exercise is largely an education process, it is best to move forward one successful step at a time. People learn best if the teaching or the training is done on a gentle gradient. If you teach them one element at a time, they will be well prepared for the final stage, which is a full-scale exercise putting into practice all that you have taught them.

These are the same steps you were working with above as "element testing." However, in element testing, your purpose was to determine the accuracy of the instructions. Now, as you "exercise" each of these elements, your purpose is to educate people and ensure they are able to follow the instructions successfully.

- *Introduce emergency invocation procedures*: The first lesson will be when you introduce them to the emergency invocation procedures and explain what those procedures are for. Once they understand the purpose, provide them with details about where they can obtain further information or support in connection with EEP. Depending on the organization and its culture, this initial introduction to the subject may take the form of an informal presentation at which they can ask questions and discuss any relevant issues. Alternatively, you might choose to send everyone a message to explain the concept and invite their questions. Another way would be to brief managers and supervisors and ask them to raise the matter at their next staff meeting. People should also have access to a copy of the EEP so that they can familiarize themselves with its contents as the exercise program moves forward.

- *Explain alarm and communication procedures*: The next step is to tell them about the alarm and communication procedures so they know what to expect when there is an emergency. They should also be told how to alert someone or raise the alarm if they suspect that something is wrong, such as a fire. This lesson incorporates a demonstration of the alarm and the communications procedures so they are able to recognize and distinguish the emergency evacuation announcement from other noises.

  One of my clients made a short video of the emergency alarm procedures which members of staff could play on their desktop computer in their own time.

- *Explore the escape routes*: Once people understand the purpose and can recognize the start of an emergency evacuation, give them a chance to explore the exit route, or routes, from their place of work or residence to the exit points. One approach to ensuring their familiarity with the routes is to encourage them to seek out the route and explore it for themselves. Another approach would be to lead them to an exit point and demonstrate how to open the emergency exit – usually a simple matter of pushing a bar to release the door. These doors are often connected to a security alarm; thus, you will need to check with the security people that it is okay to open the door before doing so.

- *Find the way to the assembly areas*: In this step, people become familiar with the escape routes which lead away from the exits towards the assembly areas. I find the easiest way to do this is to take them on a guided tour in small groups, drawing their attention to the various signs and key points of interest along the way. If the routes are pretty straightforward, you might simply allow them to check out the route on their own. The chances are that in the event of a real emergency, they would all follow the crowd. Realistically, then, you need to teach only enough of them to ensure that somebody will be there who can lead the way.

- *Become familiar with assembly areas*: Once they are familiar with the escape routes, show them the assembly areas. Point out the salient features and make sure they know where to report to the marshal who will want to check that everybody has arrived safely.

This step can be carried out successfully only by means of the guided tour approach. Be prepared to answer any questions they may have about the arrangements and procedures with regard to reporting and the ongoing communication while they are assembled.

- *Understand the stand down and return procedure:* The last of the elements is the stand down and return procedure. Explain to them who will tell them when it is safe to return to work or give them instructions about what to do next. They won't expect to stand around at the assembly area forever. There may be some caveats which you need to explain or discuss. Depending on the nature, scale, and timing of the incident there are a number of possibilities. They may be asked to go home and wait for more news, they may be told to report for work on the following day, or they may be told to go to another location. People will want to know how these decisions will be made and exactly how the next step will be communicated to them and by whom.

Finally, once you have conducted some smaller exercises in which people have practiced all the elements of the EEP, it will be time to consider running a full exercise in which people try out the whole procedure from end to end under realistic emergency conditions to prove and demonstrate that the EEPs and arrangements do actually work.

**Your regime should be a long-term, iterative plan covering the initial development, delivery, testing, and exercising as well as the ongoing maintenance program.**

## 6.3 A Delivery and Service Regime

In order to be sure that all of your planning efforts are effective and that the long-term benefits are fully realized, it is important to implement a proper means of delivering, testing, exercising, and maintaining the plans. This is where and how you turn what started as an interesting and benevolent idea into a practical reality which has the potential capacity to save many lives. This "Delivery and Service Regime"

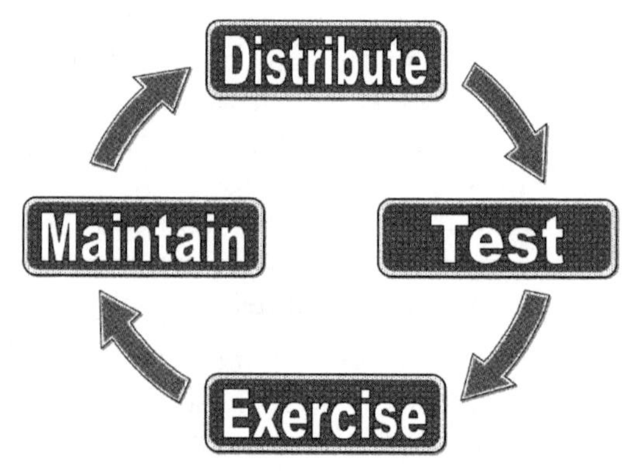

comprises four elements, and the diagram shows how it operates as an ongoing program of work.

1. Distribution of plans for review and appreciation by the users.
2. Testing of plans to check their accuracy and completeness.
3. Exercising of plans to develop competence and confidence.
4. Regular reviews and maintenance updates to ensure currentcy and effectiveness.

Map out a timetable for this regime with tentative target dates for the initiation and completion of each of these elements. Your regime should be a long-term, iterative plan covering initial development, delivery, testing, and exercising as well as the ongoing maintenance program.

All of the people who are affected or protected by the EEP program should be made aware of this regime and their involvement. It isn't necessary to give them firm dates at this point in time, but it would be helpful if you gave them some idea of the likely timeframes. For example, you might suggest that there will be two exercises a year, one in winter and one in summer. This timing will enable them to experience an evacuation under the best and the worst of weather conditions and you will be able to get feedback based on those varied conditions.

## 6.3.1 Distribution

Once the peer group review, or a similar procedure, confirms that you have what appears to be a satisfactory plans or set of plans, they need to be distributed. There are two purposes behind this procedure. The first and most obvious purpose is to provide the audience and potential users of the plans with an opportunity to review and comment upon each plan, its contents, and its apparent usefulness. This is an integral part of your built-in quality assurance process.

If you simply issue plans without calling for feedback as a part of the assurance and improvement process, it is unlikely that they will be taken seriously and will probably be parked with all the other pamphlets, brochures, and sets of instructions that people receive on a regular basis. In short, if you fail to solicit feedback, the plans will have little or no impact.

The other and more subtle purpose of distributing the plans is to raise awareness, understanding, and an initial level of confidence in preparation for the more practical stages of testing and exercising which are to follow. In order to achieve this awareness, it is essential that you manage to get everyone to take the review process seriously; you need their buy-in. It may be necessary to follow up the distribution with a subsequent request for their feedback if you have not received anything from them.

To a large extent, the distribution process will be dependent upon the scale, nature, and culture of the organization. It may also be influenced by the means of storage and the format of the plans. One approach is to store and distribute such documents in electronic format, in which case people will need to make individual choices about whether to have the plan printed out in hardcopy. This choice will also influence where and how they keep their own copy of the plan.

My preferred option is to deliver the plans personally in a face-to-face situation. When you distribute plans personally, you can engage with the recipients, and your enthusiasm can help them to realize the worth and value of being able to reach safety with a high degree of certainty no matter what the threat might be. Such distribution can take the form of a one-to-one personal handover, in which case you have to allow for the time and effort which might be required to cover the whole population. There is also the question of prioritizing the deliveries and making sure that you can get access to everybody who should be on your delivery list.

An alternative is to arrange for people to attend a workshop where you issue the plans and run through the details with the group, making sure that they fully understand and appreciate the plans and their contents. It is a good idea to ask people to sign for their copies of the plan so that you have evidence of the distribution. This also helps to get them to take the whole process seriously.

Obviously the format and overall professional appearance of the plan will need to conform to the local culture and the needs of the individuals who will be using it.

To ensure that everyone takes part in this stage of the program, you will need a tracking process to be reasonably sure that all recipients have received and read the plans. This tracking can be aligned with, or form part of, the feedback procedure. Basically, this process consists of a list of names of all the potential recipients with a space to record that they have acknowledged receipt and you have obtained their feedback. Not all the recipients will want to comment or make suggestions, but they should all be given the opportunity and time to contribute to the development and improvement process.

While feedback and comments should be welcomed at any time, you should specify a definite period for the initial contributions to be considered in connection with finalizing this first draft of the plan. All of the input should be logged and collated during this initial review period. All contributions should be acknowledged suitably to make all the people feel they are being recognized as a valued part of the process. The time allowed for the review will depend to a large extent upon the nature and size of the population concerned. If it is too short, people will feel you are trying to put pressure on them; if it is too long, people will tend to forget. You need to strike a reasonable balance. Perhaps a week or two might be about right, but you are best placed to make such judgments.

**At the end of this initial review period, go through the feedback, consider the value and implications of what people are saying, and implement any changes which are called for.**

At the end of this initial review period, go through the feedback, consider the value and implications of what people are saying, and implement any changes which are called for. The result will be the working draft version of the plan (or plans), which will then be distributed to everybody. If no significant changes were made as a result of the initial review, then you should inform all the recipients that they already have the approved version – the approval having come from their own peer group review.

## 6.4 Conducting Tests and Exercises

### 6.4.1 Testing

As discussed earlier, testing is a vital step in the development process, and each of the elements should be tested separately before embarking on a full-scale, end-to-end test. Obviously, there will be exceptions to this; where the plan and the evacuation procedures are relatively straightforward, you may feel you can embark on a full test immediately.

Whichever approach you use, make sure that the testing is properly planned and the results are documented. Your main aim is to ensure that the procedures, resources, and facilities are fit for purpose and to gain some indication of the timings involved. The documentation provides evidence of the fact that you and your colleagues have applied due diligence in connection with the welfare, health, and safety of those who are likely to be using or visiting your premises.

At the end of each significant test and all exercises, hold a debriefing immediately to give all the people involved a chance to air their views and permit you to capture any useful comments or points for improvement. Such debriefings should be a regular feature of all emergency evacuations, whether they are tests, exercises, or the real thing. It is an opportunity to share ideas and build a sense of camaraderie as well as a chance to learn and seek improvements.

## 6.4.2 The Launch Test

Once the plan has been finalized and all its elements have been proven through some form of testing, it is time to embark on the "launch stage" of the EEP program. Apart from its practical significance in demonstrating the recently developed evacuation procedures, your launch should also be a kind of cceremony – a marketing or public relations event – which marks the delivery of the completed work. Use it to remind everyone of the importance and significance of being prepared to deal with a dangerous situation, should it ever occur. There are plenty of examples you can use to reinforce the message you are trying to get across – "Safety is something we have to prepare for; we can't take it for granted. It doesn't happen automatically!"

This test is as much a demonstration as it is a test. Its principal and published objective is to check how long it will take to evacuate a representative number of people. Obviously, you will have checked and tested all of the elements in advance, and so you will know that everyone will be able to hear and recognize the alarm, follow the instructions, and reach a place of safety without difficulty.

Choose a group of people, a department, or a section of the building which you intend to invite to be part of the launch test, and check with whoever has overall responsibility for activities in that area. Explain to the responsible person what you have in mind, including the purpose of the test and the public attention which you will be trying to create as an integral part of your ongoing awareness campaign. I would expect that you will get their full and enthusiastic support, but if the person has any doubts or reservations, then I would suggest your response should be: "Okay, we can try somewhere else." It is most unlikely that anyone would want to resist this opportunity, but it is always best for you to prepare an alternate strategy. Once you know how long it takes for this number of people to evacuate the premises and reach the emergency assembly area, it is relatively straightforward to estimate the time it will take for a complete evacuation. (Details of how to carry out such a procedure are explained in the United Grand Lodge of England case study later in this section.)

To ensure that you obtain the measurements for your calculations, make sure that a couple of marshals or observers are in position before the start of the test with stop watches ready to time the whole proceedings. Make sure that these people are able to hear the initial alarm so they know when to start timing. If you have any doubt about their ability to recognize the alarm from where they are positioned, arrange for some form of signaling arrangement to alert them. The easiest way is often to use mobile phones to keep them advised of what is happening. In some situations, it might be possible to use a visual signal such as waving or dropping a flag. One client of mine actually uses a smoke signal to act as the trigger for the alarm and as a signal to the observers that the test was under way. They adopted the smoke signal idea as the trigger for all of their subsequent exercises; thus, even their neighbors are familiar with their EEP program of regular evacuation exercises. Their local emergency planning officer regards them as model citizens.

If the exit routes or evacuation paths are anything other than straightforward, arrange for observers to be present at any of the decision points or bottleneck points along the way so they can report back on how the people coped. During the subsequent debriefing, you should also give the participants a chance to comment on any difficulties or problems which they experienced.

Make the results of the test known to everyone together with the calculations regarding total evacuation time. Remember to make it clear that the total evacuation time is an estimate based on the assumption that a larger crowd would act in the same manner and that larger volumes would not lead to unexpected difficulties in the evacuation procedure. This clarification is important because it leads naturally on to the necessity for a full exercise to confirm the prediction that the procedures you have in place will ensure that everyone can reach safety within a reasonable period of time.

**...for this first demonstration of your emergency evacuation procedures, check whether there is any relevant legislation, regulation, or guidelines regarding the frequency or importance of such exercises.**

## 6.4.3 Exercising

Everyone was made aware of the launch test, which was promoted as a significant event in the development and delivery of the EEP program. The first full-scale exercise warrants the same, or even more, publicity. It is an ideal opportunity to reinforce the need for, and the benefits of, your EEP program.

Before making the announcement and final preparations for this first demonstration of your emergency evacuation procedures, check whether there is any relevant legislation, regulation, or guidelines regarding the frequency or importance of such exercises. You may identify regulations which apply locally, that are industry-specific, or that apply to your particular group of people. If you discover any rules that need to be taken into account, then you should make sure that everyone is made aware of them to accentuate the importance of what you are doing.

Because it is the first event of its kind, you need to give plenty of notice to all parties so that everybody is properly prepared. Depending on the audience and the complexity of the plans and procedures, invite everybody to a briefing meeting or perhaps simply ask them to read through their plans in order to make sure they fully understand what is expected of them. In some situations you may even arrange an escorted walkthrough for some members of your group prior to the actual exercise.

If your plan calls for, or allows, marshals to assist, direct, or guide the evacuees, then ensure they are deployed before the start of the actual evacuation so that they are already in position when the crowd begins to respond to the first signal. If they have high visibility vests or jackets they should put them on when they are first deployed – people will get to see and recognize them as they get ready and move into position. Having the marshals in place and ready for the start will ensure that it all happens in an orderly controlled manner. Once everybody is familiar with the procedure and the marshals know where to place themselves to the best advantage, then you can run a more realistic version of the exercise in which everybody starts at the same time.

While you should try to ensure that this important initial exercise is a public display, be careful not to turn it into a pantomime. It should be experienced, and consequently remembered, as a serious educational event rather than a frivolous entertainment.

From the perspective of timing, ensure that your exercise doesn't coincide or conflict with any other important occasion within the organization or the community. This will ensure that attention is not diverted or split between the two. Avoid any peak periods when an exercise might be regarded as an unnecessary and inconvenient disruption. On the other hand it isn't necessary to take pains to ensure that every single person is available; aim to involve the majority rather than the entirety of your population. Those who miss this exercise will catch up at the next one. If a real incident should occur in the meantime, they will be able to follow the lead of others who will have previous experience.

Monitor the performance of all those who take part and let everybody know what was learned during the exercise; inform them of any planned changes or improvements which the event has highlighted. Any changes should be discussed at the debriefing and comments noted for your report. Hopefully, it will all have run smoothly and no changes will be required. In this case the overall evacuation time will be the main piece of information for your report.

In any case, your report should give an indication of the likely date, or at least the timeframe, for the next one.

The marshal and observers should be asked to share their views about the plans, procedures, and their interpretation during the debriefing, but you should discourage them from making any negative comments about the performance or capabilities of any specific individuals; any criticisms should be expressed in very general terms. If you permit criticism of the performance of specific individuals, you face a distinct danger of a debriefing that does more harm than good. I would be happy to hear this: "Some of the signs or the instructions were not very clear, which probably caused some people to take the wrong direction." I would not want to hear comments like this: "Charlie seemed to ignore all the signs," or "The accounts team made loads of mistakes and ended up in the wrong place."

Ideally, every exercise should be arranged and managed in such a manner that the result can be seen as a success and the participants congratulated on their performance and positive attitude. If the result is anything less than this, then it is the development and delivery of the exercise which is at fault – not the participants. I always like to end an exercise with a few words of praise such as: "Despite all the little problems we threw at you, you did very well. Congratulations!"

You may choose to introduce a few additional problems for the participants to deal with when taking part in an exercise, especially when you believe that the majority of the people have developed a reasonable level of competence. This will make it more interesting and stretch the minds of all those involved. However, you should refrain from any attempt to complicate or confuse the evacuation proceedings. Extra challenges can be very useful in those situations in which we are trying to get the players to think on their feet and prepare to tackle all sorts of difficulties under stressful circumstances. In a crisis management or business continuity exercise, for example, such unexpected occurrences would add to the realism and provide additional learning opportunities; however, introducing the unexpected has no place in an emergency evacuation exercise where the aim is to develop confidence in, and familiarity with, the prescribed procedures. This is not the time and place to start playing games or to introduce unnecessary complexities.

## 6.5 Review and Update

The review and update stage of your EEP program is in many ways the simplest but also the most arduous. Theoretically, it is just a matter of keeping track of what is going on and making sure that the plans and arrangements are in line with the needs and expectations of those who occupy or visit the premises. In practice, this stage consists of a series of routines which are easy to ignore, forget, or neglect. It takes a special mind set to pay attention to the administrative detail on a long-term basis. The thinking here is very different from that which is required when you are setting out to explore and develop something new. Those early stages are a series of projects with defined outcomes and timeframes; as each deliverable is complete, one gets a sense of achievement which is quite satisfying. Now you are facing a long-term process with few significant milestones. Any satisfaction comes from knowing that everything is more or less up to date.

**All of the work done during this stage should be seen to conform to the required standards or guidelines as described in the EEP policy statement.**

If the review and update process is neglected, you invite the distinct danger that you will be putting people's lives in jeopardy; you will certainly not be demonstrating that you are following the EEP policies and official procedures. All of the work done during this stage should be seen to conform to the required standards or guidelines as described in the EEP policy statement; thus, it needs to be documented properly with a *full audit trail*. There's no rocket science involved in this part of the job, but it is necessary and worthwhile.

*Document control procedures* should be followed to ensure that everyone is referring to the correct or latest version of the plans and any other related documents. Follow the version control system which is already in place within your organization; there is no need to reinvent the wheel. The minimum requirement here would be to include the date and version number on every document. Page numbering should be required for any multi-page document. Additional information such as ownership, authorization, approval, copy number, distribution, etc., is not essential in this connection, but may be an obligatory element of compliance with your internal control systems. Wherever there is an option, try to keep the plans as simple as possible.

Keep a record of all the document control characteristics together with details of the review and update process in order to ensure that you follow the routine. To keep such a record means you need to set up and maintain an *administration log* of some sort in which you will record all such information. This log should be supported by a rolling action plan which provides a timetable indicating when reviews, updates, tests, and exercises are expected. Where there is more than one current version or type of evacuation plan to satisfy the differing needs of a varied or widespread population, your schedule of work will need to take all of the variants into account. If you are using a spreadsheet for these records, it might be convenient to have a separate sheet to keep track of each specific plan or group of plans.

Apart from the action plan, which gives an indication of what needs to be done and when, keep an *EEP plan status report* to track any changes or improvements which have been implemented as a result of the ongoing review process. Such a report will give you, and others, an instant overview of the EEP program and its effectiveness.

It is also important that you get your sponsor's approval or endorsement of this ongoing schedule of work. Apart from being politically and diplomatically correct, it will help to reinforce the good working relationships which you will have developed. Your sponsor should also be given access to the status report; this will give your sponsor the assurance that the EEP program is meeting its objectives.

## 6.6 "Smart" Fire Drills

> ***"Fire and related gas build-up caused explosions in the sewer lines beneath the city's streets. The explosions were so powerful that they sent manhole covers flying into the air and shook nearby buildings. Evacuees left the relative safety of the buildings for the dangers of the street and flying manhole covers, each weighing more than 100 pounds."***
>
> John Glenn, BCM practitioner

You can add value to conventional fire drills – making them "smart" fire drills – at very little additional cost, by considering the evacuation and communication procedures in the light of lessons learned from actual experience.

In this section, we introduce the concept of conducting a *fire exposure analysis (*FEA), which merges some of the concepts of business impact analysis (BIA) with routine fire drills. While almost everyone is familiar with the concept of a fire drill, the BIA process is a rather more specialized tool from the world of *business continuity management* (BCM). The resulting FEA procedure is an expedient tool which can provide you with a powerful argument for development of sound contingency plans and the investment of time, money, and effort in your emergency evacuation plans and procedures.

Bear in mind that any bid for additional budget to fund improvements to the evacuation procedures, resources, and supporting facilities is not likely to result in an immediate burst of funds. Usually it takes several months, if not a year or more, to obtain full support and funding for any project that appears to lack an obvious, direct return on investment. A cost/benefit analysis or a fire impact analysis will help your argument.

All I, or anyone else, can offer is a few hints and tips. The rest is up to your own persistence, wisdom, and persuasiveness.

***The Standard Fire Drill***: Regular fire drills are an established activity in most business premises. However, they are relatively costly exercises which bear little fruit. People are interrupted in the middle of their work, and their minds turn to other things. Since there is no rush to return to their place of work, they are provided with an unexpected opportunity to socialize for a while.

On the way back, they tend to congregate around the coffee machine where the socializing continues. It is often half an hour or more before the workforce is fully re-engaged in its work. Yes, you do establish that everybody can hear the alarm and get safely out of the building. But surely, you have an opportunity here to learn a lot more without significantly increasing the cost and inconvenience. Thus, I would suggest adopting a forward-thinking approach to the evacuation procedure itself.

**There are many lessons to be learned from almost any serious fire, some quite specific but many generic enough to apply to most cases.**

## 6.6.1 On The Way Out

It makes sense to advise all of your staff to ensure they pick up and take their personal possessions such as keys, jackets, and handbags with them, providing they are at or near their desks. When people have their possessions with them, they will be less inclined to rush or panic, and they will be better prepared for what comes next. That could be "return to work," "stand around," or "go home." Standing around without a jacket can be unpleasant if it is cold, wet, or windy. Going home without any keys or money will be difficult, especially if people have a fair distance to travel.

There are many lessons to be learned from almost any serious fire, some quite specific but many generic enough to apply to most cases. In the UK, we certainly picked up a few learning points from the Digital Equipment Corporation (DEC) fire in 1989.[1]

The company had recently moved into a newly built leased building in Basingstoke, Hampshire. Several hundred staff members were busy at their desks when the fire alarm sounded. Fortunately, a fire drill had taken place recently, and so they were all familiar with what to do and thought the alarm to be no more than yet another drill. They were wrong!

Construction work in the roof-space of the three-story building had set fire accidentally to some combustible material. The strong wind blowing at the time caused the fire to develop quickly, and within a few minutes the entire roof area was ablaze.

Although everybody got out fairly quickly, the problems were not over. As the flames began to devour the entire structure, employees watched the burning roof collapse on their cars which were parked neatly around the building. Then they realized that their jackets and handbags containing railway tickets,

[1] This anecdote was provided by Colin Ive, MBA, MBCI, who was both an employee and a part-time firefighter at the time.

car keys, house keys, credit cards, wallets, and other personal items were still in the building and effectively lost to the fire.

Unfortunately, the situation was aggravated by the fact that, as a result of a request by a senior manager of the company to the fire service, the fire sprinkler system had been turned off. Incidentally, this led to the owners of the building setting legal history when they successfully sued the fire service for negligence and were awarded £16 million compensation.

Fortunately, the culture of DEC was very people-focused, and the HR department arranged to meet the needs of staff members for taxis, locksmiths, cash, etc. This good example of crisis management was compounded by the company's ability to relocate the staff very quickly and reinstall the mainframe computers in nearby alternative premises. Luckily, the equipment had not been damaged by water from the inactive sprinklers. The overall speed of recovery so impressed many of DEC's customers that the business realized they could use this experience as an excellent example of disaster recovery, and DEC's disaster recovery service was born.

The central theme here is how an evacuation scenario might develop once the people are clear of the building. A number of possibilities need to be taken into account. You need to ensure that your people are prepared to receive and carry out whatever instructions are appropriate at the time.

## 6.6.2 While They Are Out

Ask managers to deliver a message or instruction to their staff once they reach the assembly point, which is precisely the sort of thing that might be called for in a real emergency. Perhaps the message could be to get everybody from each group or department to do something in particular before they return to work. Ideally, all managers would be able to direct every one of their staff to carry out a simple task. However, I suspect many would fail.

Consideration should be given to the question of relaying messages and instructions when selecting assembly areas and muster points. The ability to communicate with staff will depend on the quality of the procedures, how well people understand them, and whether they are inclined to adhere to them. This is the point at which training and practice help to achieve a reliably effective response to a fire alarm.

Think about those occasions when it may be necessary to evacuate the premises for other types of threats to health and safety. Different emergencies may require the consideration of alternative signals, exit routes, and exit points as well as alternate assembly areas.

### 6.6.3 Fire Exposure Analysis

When the people do eventually return to work, require each person to make a fire exposure list, i.e., what would have been lost in a fire. Combine all of the lists from a section or department to get an idea of your overall vulnerability. The accumulated potential losses might highlight the need for better protection or more precaution.

***Fire Exposure Analysis Part 1 – Regular Fire Drill***: A fire exposure analysis might start off with a regular fire drill to set the scene before entering into the data collection and analysis stage. Although they are described here as two phases of the same procedure, these two phases can be conducted quite independently. For example, you can conduct the analysis as a standalone, desktop activity.

- **Step One – Sound the Fire Alarm**

  Simply sounding the fire alarm establishes that the alarm system works, everybody can hear it, and they recognize what it is. You have proved that it is effective.

- **Step Two – Evacuate the Premises**

  Evacuation demonstrates that all the people know what to do and they are able to escape in the event of a fire. You can monitor the evacuation to ensure that everybody does get out of the building.

  It is also common practice to record the overall evacuation time as an indication of the effectiveness of the whole process. This figure may influence the urgency of a real-life response to an incident. If we think we can get everybody out quickly, we may delay our alarm. On the other hand, if we are worried that the escape time could prove rather lengthy, we may be tempted to trigger an alarm without fully investigating the cause and its potential effects.

- **Step Three – Roll Call**

  Conducting a roll call is how you check that everybody is safe and accounted for. At the same time, you can make sure that the assembly point is suitable and known to everybody. In a real emergency, it is quite likely that the roll call will not be able to account for everyone, especially if strangers or visitors are in the building at the time.

- **Step Four – Return to Work**

  Finally, once you are sure that everybody is safe and accounted for, you normally allow them to return to their place of work and carry on as normal.

You might time the evacuation to check how long it takes to clear the building. You might also check how much actual working time is lost in such an event, since a considerable amount of time may elapse before the complete working routine is fully restored.

For a simple fire drill, this is the end of the exercise. However, we suggest that you carry out a second phase devoted to an impact analysis based on the fire scenario.

**This step is designed to put a realistic value on the losses that might accrue in the event of an office fire.**

*Fire Exposure Analysis Part 2* –Assess Potential Losses: This step is designed to put a realistic value on the losses that might accrue in the event of an office fire. The principal, irretrievable losses would be those associated with missing paperwork which was in use at the time of the fire. In a factory, retail, or service environment, valuable goods and materials could be at risk in the event of a fire. The technique needs to be tailored to suit the operational environment in which it is applied.

- **Step One – Work Status Records**

  Employees should be provided with a check sheet to record their work status at the time of the fire alarm or other incident. You want them to note what they were doing at the time and the documents with which they had been working. The point is to establish what would have been lost in the event of a real fire. Work in hand is likely to remain incomplete until the building can be reoccupied, and loose documents are likely to have been lost forever, meaning a loss of any potential work which would use those paper documents as reference material. For the purpose of the exercise, assume the building will remain out of bounds for three weeks or more. If it were a serious fire, it could take many weeks before the building is returned to service. Allow a further week to deal with the inevitable backlog of work and to reestablish normal operations. In other words, consider the impact of four weeks inaction, by which time a considerable backlog of work is likely to have built up.

  A *pro forma* FEA Work Sheet is included in the accompanying EEP Toolkit.

- **Step Two – Collate Work Losses**

  Within each business unit, someone should be tasked with collating the individual work status records to determine the full loss potential or exposure at the local level. Wherever possible, the cost of loss should be expressed in terms of the ultimate gross value, i.e., the final market value of the outcome of each piece of work. The output from this exercise will be a realistic estimate of the amount of actual and potential work lost within a business unit over the first four weeks after a fire. Most of this loss is probably going to be irrecoverable.

- **Step Three – Compile a Total Cost**

  Once the potential work losses have been collated within each business unit, you can start to build the full picture. By combining the losses across the various business units, you can now compile a total cost estimate, which will give an indication of the true cost of a major fire at that location. However, this exercise takes only a relatively short-term view covering a period of only a few weeks, whereas many business relationships, processes, and projects are measured in terms of months or years.

- **Step Four – Consider the Full Impact**

  In the longer term, future business may well be affected by an interruption to normal operations. How many customers will go elsewhere? They may lose faith and seek alternate suppliers, or they may simply decide not to proceed if their key supplier is in trouble. Rumors will abound, and competitors may use the opportunity to tarnish your image.

  Obviously, insurance will be seen as some form of counter-measure, but it is hardly likely to cover the full impact on future business. In fact, it may not cover more than a fraction of the immediate losses.

***The Importance of Timing***: When embarking on this type of adventure, you must avoid imposing additional risk or strain to the normal working operations of the business. Try to choose a relatively quiet season of the year so everyone can afford to spend the time on what is, after all, a peripheral activity for most of them. If you were to attempt this type of exercise at a peak period, you could very well generate resentment and fail to get the level of support you were hoping for.

### 6.6.4 Available Safe Egress Time (ASET) and Required Safe Egress Time (RSET)

The design philosophy behind many modern buildings involves the use of large open-plan areas which allow freedom of movement and a sense of spaciousness. However, such areas can, and often do, make it rather more difficult to contain, prevent, or extinguish fires, which is why architects tend to work closely with fire safety engineers when designing public or commercial buildings. The use of materials known to limit the size and spread of a fire is a key aspect of passive fire protection, which aims to allow people to reach safety before the fire and smoke overwhelms them.

Fire safety engineers bring with them a wealth of knowledge and a number of tools and techniques to develop a safer working environment for the users and occupants of any building. Much of their knowledge and skill is based upon the vast amount of fire research which has been conducted over the years and the innumerable case studies which are available to them.

Engineers commonly work with two related concepts to analyze the conditions inside a building in the event of a fire: the *available time* is compared with the *required time* in order to determine whether a building can be evacuated safely. While the original concept was to be able to verify the safety of occupants in the specific case of a fire, the same principles apply to any evacuation, no matter what the cause. After all, passive fire protection is primarily concerned with enabling people to reach safety, and that remains the common purpose of all evacuation procedures.

The fire safety engineer aims to show that the time available before conditions become untenable due to the effects of flames, heat, or smoke should always exceed the time required to safely evacuate the building. This margin of time enables engineers to provide assurance that safe evacuation of a building will be possible.

Fire safety engineers use the terms *available safe egress time* (ASET) and *required safe egress time* (RSET).

***Available Safe Egress Time (ASET)***: ASET is normally defined by a set of acceptable criteria such as temperature, density, and toxicity of smoke within the areas which might comprise the exit or escape routes.

Of course, you can modify the ASET figure to take into account many factors which are specific to a particular site or location. For example:

- The type and quantity of flammable material, otherwise known as the fire load.
- Ceiling heights and shapes.
- Smoke ventilation systems.
- Smoke and fire barriers.
- The shape and layout of the rooms.

The fire safety engineer uses these factors to calculate the ASET for each specific building or space, always erring on the conservative side because lives might be at risk as a result of their recommendations.

Traditionally, these assessments were calculated using pen and paper, but today, most fire safety engineers make use of computer modeling to provide visual representations. Two basic types of computer models are used for the development of fire scenarios: the *zone model* and the *computational fluid dynamics* (CFD) model.

**Traditionally, these assessments were calculated using pen and paper, but today, most fire safety engineers make use of computer modeling to provide visual representations.**

Simple zone models are used to predict the height of smoke layers and viable conditions in buildings or rooms with a straightforward layout.

Where the situations are more complex, CFD models are able to provide a more realistic assessment of the changing conditions as a fire develops and the smoke spreads. CFD models take into account factors like fire growth, flame spread, suppression, detection, sprinkler activation, and other related phenomena to give an accurate prediction of the likely consequences of a fire.

***Required Safe Egress Time (RSET)***: When quantifying the RSET, three components have to be taken into consideration:

- Detection time.
- Pre-movement time.
- Movement time.

**Detection Time:** *Detection time* is the time from the fire starting to the moment when an occupant becomes aware of it, ending when an alarm or warning signal has been raised.

The detection time, and therefore the RSET, depends to a large extent on the type of detection system in use. For example, manual fire alarm systems tend to produce a longer detection time than automated systems with detectors throughout the building.

**Pre-movement Time:** The *pre-movement time* is the amount of time taken by occupants to recognize the danger signals and react to the fire alarm. Recognition includes the time taken for occupants to take the alarm signals seriously and realize they need to respond to the emergency situation.

Their initial response may include a number of activities, such as trying to find the source of the fire; shutting down machinery; warning other people; protecting exposed valuables, such as money and documents; gathering children and other family members; assisting colleagues, especially those who are disabled, handicapped or injured; finding their way; and fighting the fire.

Pre-movement time can vary because it is subject to so many variables, including whether it is a managed or an unmanaged evacuation. In a managed evacuation, the level of assistance can be a major factor as can the amount of safety training, the type of occupants, and whether the building is fitted with a public address system that can be used to provide instructions in an emergency situation.

**Movement Time:** *Movement time* is the time taken for the occupants of a building to reach safety once the actual evacuation has begun. Reasonably accurate predictions of evacuation movement patterns during an emergency can be made in several ways. These predictable flow patterns have led to the development of egress models which can both test the evacuation arrangements and predict movement times, or evacuation times, with a fair degree of precision.

Obviously, your overall movement time is dependent upon a number of variables, such as the number and location of available escape routes; the widths of escape routes and exits, which often form "bottlenecks;" and the nature and number of occupants, their walking speed, and their dependency on the support of others or special equipment such as wheel chairs.

Due to both the complexity of the evacuation process and the number of factors influencing the actual movement time, manual calculation methods are suitable only for assessing relatively simple escape arrangements.

To model the evacuations from more complex buildings, fire safety engineers would normally use computer modeling tools.

Evacuation models are able to take account of all these parameters, calculate a potential evacuation scenario, and give information about the use of escape routes. They can also predict accurately the time required to move from a certain area to a place of safety, generating data which can be used to make sound assumptions for the overall movement time under a variety of circumstances, such as the worst case scenario.

The sum of detection time, pre-movement time, and movement time results in the RSET for the building. A safety margin should be added to this theoretical figure to provide a realistic RSET figure to be used for design purposes.

The bottom line is that safe evacuation can be assured if the assessed ASET exceeds the RSET by a reasonable margin.

## 6.7 Case Study: United Grand Lodge of England

The following case study looks closely at emergency evacuation planning carried out by a progressive business continuity manager working in a unique environment and supported by an imaginative group of executives.

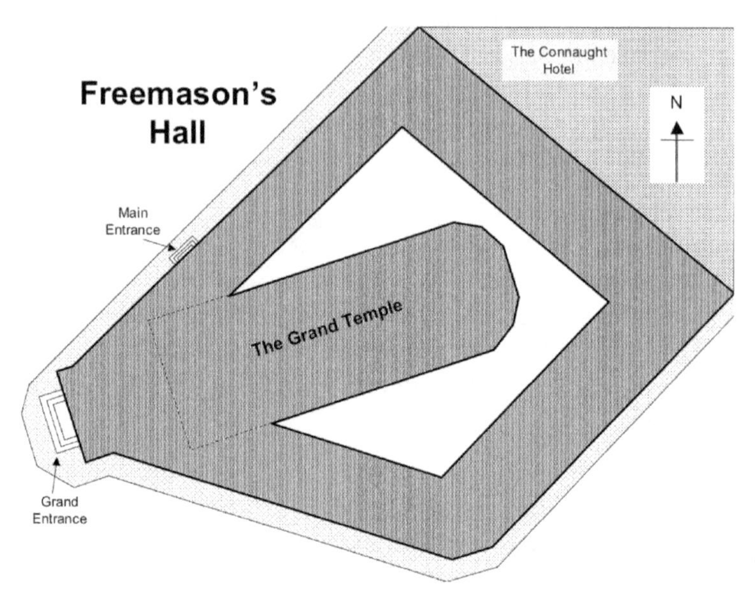


Peter Jack comes from a disciplined high-tech back ground, working as a technical officer in the Royal Navy before he joined the United Grand Lodge to manage its IT. Gradually, over time, he was given other tasks and expanded his responsibilities, which eventually included aspects of security and business continuity. Whenever he starts on a project in which he is likely to get out of his depth, he calls someone in to give him advice and support while he is learning the tricks of the new trade. He is a quick learner, and this approach has always stood him in good stead, which is perhaps why the company keeps stretching his boundaries.

The United Grand Lodge of England (UGLE) is the main governing body of Freemasonry within England and Wales and districts overseas. It is the oldest Grand Lodge in the world, with a history going back to 1717. The headquarters building is known as Freemasons' Hall, located close to Covent Garden in central London. It is an imposing art deco building, covering almost nine acres (404,000 sq ft) of accommodation.

Freemasons' Hall is a "Grade II" listed historic building, i.e., "a particularly important building of more than special interest." This designation means it comes under severe protective restrictions and may not be demolished, extended, or altered without special permission from the planning authorities.

Within the Hall are a number of discrete areas; the principal area is the Grand Temple which can hold up to 1,600 people. In addition, it contains 23 other Masonic temples with capacities from 25 up to 450 people. These are all in constant demand by lodges and chapters which hold regular meetings there. The building also houses a shop, a library, and a museum as well as the four major Masonic charities and the administration to support these various operations. During the week, the building is open to the general public for guided tours of the main features of the Hall.

Freemasons' Hall is unique in its layout. Because the site is a rather strange shape and doesn't align with the compass, the Grand Temple was built on a diagonal to the rest of the structure in an attempt to have it facing, more or less, towards the East, in common with most Christian places of worship. This unusual design makes it very easy for a new or infrequent visitor to get lost or disoriented within the complex building.

With so many visitors – many of whom are elderly, frail, or disabled – emergency evacuation planning in such a complex building is a considerable challenge. Additionally, the structure must be considered as an iconic building, and hence a potential target for terrorism.

When I first met Peter, he was looking for help with his business continuity plan, but it soon emerged that evacuation planning was rather more important for the United Grand Lodge. By its very nature, the business would continue, no matter what interruption might occur. However, the health and safety of the many thousands of visitors who visit this prestigious building regularly was, by comparison, a matter of primary importance and concern.

***How long will it take us to empty the building?*** The principal question was: "How long will it take us to empty the building?" Clearly, there was no chance of carrying out a live test with a capacity audience gathered together in the Grand Temple, all attired in their full Masonic regalia. The normal exit time for such a large assembly was anything up to an hour as they gradually changed into their street clothes and withdrew from the building, perhaps calling into the shop or visiting the museum on their way out.

My inclination was to develop a mathematical way of estimating the evacuation time based upon a limited, controlled exercise using a small group of volunteers.

Working from the premise that egress would be limited by the throughput or flow rate at the slowest point – that is, the narrowest or most difficult section, of the escape route – I came up with the following method of estimating the overall evacuation time.

Below, our case study explores a practical example of estimating evacuation time in the field. At first we used a rather simple mathematical model, and later we were able to confirm our conclusions using a sophisticated computer based modeling tool.

## 6.7.1 Hypothesis

The total time taken to evacuate a crowd from a venue is the sum of travel time plus crowd time, where *travel time* is defined as the amount of time taken for the average person to exit the venue and reach a position where he or she can be reckoned to be safe. *Crowd time* is the total time taken for the rest of the crowd who are inside the venue to follow through and join the others within the safety zone or assembly area.

We are using the term *venue* here to indicate a space in which large numbers of people are likely to gather, such as a meeting room, a lecture theatre, or an auditorium. It is assumed that the venue is contained within a larger structure or complex which we are referring to as the *building*. The path from the venue to the point of exit from the building, called the exit, is referred to as the *exit route*. The path from the exit to the area of safety or assembly area is known as the *escape route*.

Crowd time, or the total evacuation time, is reckoned to be the flow rate multiplied by the crowd size where flow rate is defined as the number of persons per unit of time. The overall flow rate is controlled or determined at the *pinch point*, the narrowest or most difficult part of the exit route. Crowd size is taken to be the maximum number of persons likely to be inside the venue at the time of the evacuation.

**Crowd Time = Flow Rate x Crowd Size**

## 6.7.2 An Estimation Procedure

***Data Capture***: In order to capture some realistic data upon which to base our calculations, it is necessary to simulate the logistics of an evacuation. This simulation can be accomplished by arranging for a small crowd, of say 100 people, to take part in an evacuation exercise. In our calculations, we will refer to this crowd as a *volume*.

At the emergency assembly area, which we will refer to as the place of safety, we measure the time taken for the first person to respond to the alarm or a start signal, exit route, reach the exit via the exit route, travel along the escape route, and enter into the place of safety. The total time taken for this journey should be recorded as the travel time. For example, it may take someone 5 minutes to emerge from the venue and reach safety.

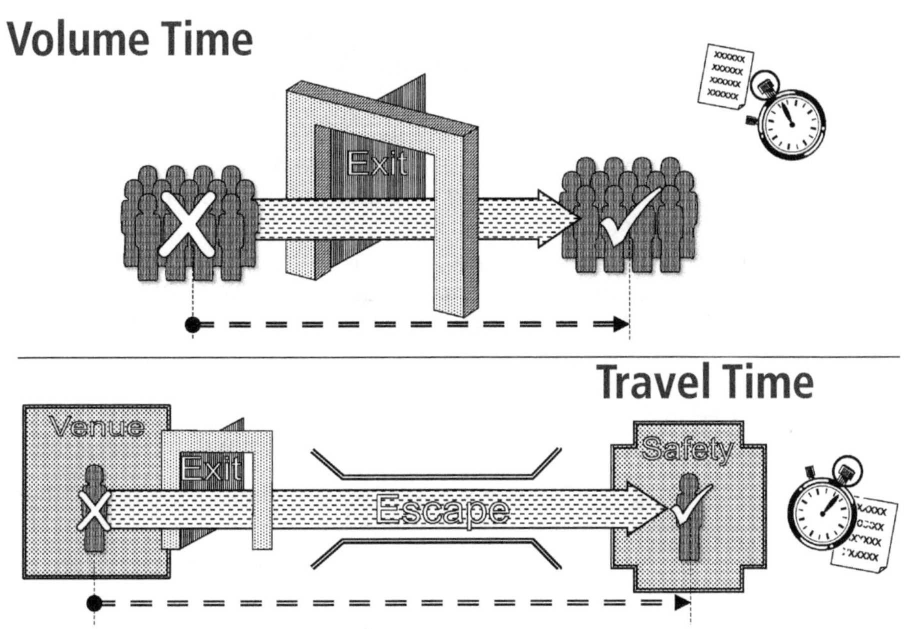


Measure the time taken for the rest of the crowd, or the volume, to follow the first person. Record this as *volume time*.

***Calculation:*** Divide the total volume time by the actual volume to determine the average individual flow rate. This gives us the average time taken for each person to pass through the slowest point in the evacuation procedure. For example, if a crowd of 100 people should take 10 minutes to reach safety, the flow rate would be:

5 minutes / 100 = a flow rate of one person every 3 seconds

Then if we wish to know the crowd time for a much larger group, say 1,000, we can calculate the time as volume times flow rate like this:

**Crowd Time = Volume x Flow Rate**
**for 1,000 people**
**=**
**1,000 x 3 seconds = 3,000 seconds or 50 minutes**

To this figure, we need to add the travel time, which is how long it takes these people to complete the full journey of the evacuation procedure. We assume that they will all travel at the same average speed, i.e., normal walking pace.

In our example, we arrive at a total evacuation time like this:

**Total Evacuation Time = Travel Time plus Crowd Time**
**=**
**5 + 50 minutes = 55 minutes**

Of course these are purely hypothetical figures. We would hope that the flow rate from any significant venue would be considerably higher than one person every 3 seconds. Indeed, it might be more like one person every half second; in other words a total evacuation time for 1,000 people of $(1{,}000 \times 0.5 \div 60) + 5 = 13$ minutes and 20 seconds. In practice, we would probably want to round this figure up to 15 minutes and to make sure that the exit route from the venue to the exit was safe for a minimum of 20 minutes. We would probably aim to ensure that the escape route from the building to the safety of the assembly area was protected from danger for at least 30 minutes. Such protection would be afforded by safety measures like fireproofing and the provision of covered walkways.

***Congestion Points:*** Inevitably, within most buildings, there will be more venues or places where people are normally gathered than there will be exit points.

Therefore, it is important for us to realize, and take account of the fact, that exit routes will converge and become more crowded as they approach the exit. In a typical multi-story building, this crowding happens on both the horizontal and vertical planes as shown in the figure below. Initial exit routes are shown as dotted lines and converged exit routes are shown as solid lines.

Ideally, the architecture should allow for the exit routes to become wider as they merge but, in practice, this is sometimes difficult to achieve or maintain, especially in older or reconfigured buildings.

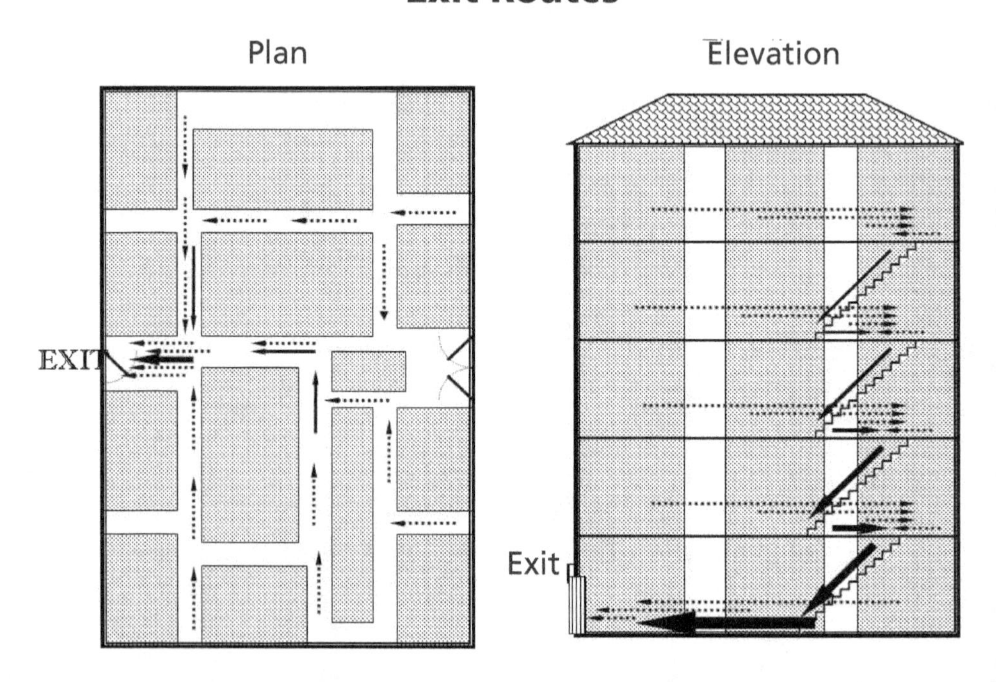

***Further Developments:*** This estimation technique is a relatively simple, approach which can provide the user with a crude approximation of the time taken for a crowd of people to evacuate a venue and reach a place of comparative safety. For a more accurate, realistic picture of the movements of people in an emergency evacuation, a number of software tools are on the market to develop useful models.

At United Grand Lodge, Peter was given the opportunity to explore some more advanced approaches to evacuation planning through the use of software which was developed by the Fire Safety Engineering Group at the nearby University of Greenwich. EXODUS, a computerized modeling tool, allows the user to simulate realistic interactions between people, fire hazards, and the infrastructure. It provides an accurate means of predicting the outcomes of various scenarios and comparing the results in order to create optimum solutions without the need for cumbersome people-based trials.

Before he was able to make use of the software effectively, he worked with the developers and his architects to build accurate models of the building. Then he was able to change the virtual structure of the building and experiment with various ideas for alternate configurations and associated evacuation procedures which might be considered. Through this experimental work, he was able to come up with solutions to his evacuation problems that were a radical departure from the norm. More importantly, he was able to prove the viability of his new ideas to both his directors and the fire safety inspector. He needed the support of his directors to fund the engineering involved, and he required the approval of the fire safety inspector to comply with health and safety legislation.

After a great deal of thought and discussion, he was able to implement a unique set of solutions to the problems presented by the building and its regular customer set. Each of these ideas had also to take account of the fact that this was listed as a historic building because of its architecture, and thus any alterations had to be consistent with the art deco style.

*A User-Friendly Solution*: The final solution was to treat the United Grand Lodge complex as a set of adjacent buildings rather than lump all the venues into one single evacuation procedure. By treating these areas as separate entities, it was possible to view each of them as a temporary safe space in the event of fire, or other incident, occurring in a neighboring sector. Such a strategy depended upon being able to isolate each of these spaces and also to ensure that people would be able to move safely and easily from one space to another. In principle, then, this was an *invacuation* rather than an evacuation approach.

Since many of the people visiting Freemasons' Hall were likely to be elderly, frail or disabled, the term "safely and easily" meant avoiding stairs.

The requisite isolation was achieved by a set of electronic controls which would ensure that all of the existing fire doors throughout the corridors of the building would close automatically in response to a fire alarm. The system uses an acoustically triggered device to hold each of the doors open and to release the spring-loaded doors whenever the controls detect the specific sound of the fire alarm.

Access between sectors was assured by a combination of methods: redesigning the exit routes, adjusting door openings to suit the revised routes, and providing suspended walkways between the northern and southern blocks. These walkways, or pedestrian bridges, had to be fully enclosed through the installation of a roof and sidewalls to provide adequate protection from smoke and fire, and they also had to be sited and constructed to comply with both the regulations governing a listed building and the requirements of the Health and Safety Executive.

## 6.7.3 The Final Plan

The final evacuation plan includes a number of options which reflect the various possible scenarios that might trigger an evacuation. Obviously, the plan includes a comprehensive set of decision criteria to ensure that the right choice will be made according to the prevailing circumstances. Generally, the priority is to clear the affected area, or areas, as quickly and smoothly as possible. Where and when it is practical, visitors will normally escape into a designated safe space within the building. Once they are all clear of the danger, arrangements can then be made to transfer them to a normal external assembly area. This second phase of the withdrawal can be carried out at a relatively leisurely pace as these spaces are likely to remain safe and relatively comfortable for quite some time.

This concept of phased emergency evacuation is an interesting development, although it might not be appropriate for most regular commercial buildings.

*The above account of the Grand Lodge of England is based on the real experience of this organization and is recounted with the permission of Peter Jack.*

# 6.8 Emergency Notification

So far, throughout the text we have assumed that the key message in any emergency is "Everybody Out" and it should be communicated loud and clear to everybody. This assumption is predicated upon the underlying assumptions that appropriate signaling protocols and communication systems are already in place. It also presupposes that it is both necessary and advisable to grab everyone's attention and alert them to the danger.

Before you finalize your emergency evacuation plans and begin to practice them, let us review these concepts and consider the potential gaps in our thinking so far.

## 6.8.1 Signaling Protocol

Prior to sending out a signal of any sort, you must work out who your target audience is, where they might be, their state of mind, and their current situation. At the same time, it is important to take account of the same factors in regard to others who may be interested in, or affected by, the information you are putting out.

There are four types of message which you must deliver to those who may be in danger on, or near, your premises.

**General Guidance** is the background information regarding exit routes, exit points, and escape paths. Most of this will be provided well in advance and take the form of signs and plans which should be reinforced through a training and exercise program. The purpose is to prepare people to be able to find their way to safety whenever their health or safety might be threatened. You also need to ensure that these messages are delivered in a way that is easily understood by all members of your audience.

This general guidance should be made available at all times to anyone who enters your premises, irrespective of their intentions or the duration of their stay. It should also include details of the invocation procedure so they know what to expect and how they should react. This is not just good citizenship; it is, in most jurisdictions, a legal requirement.

**A Warning or Alert** is the signal to your audience to prepare to leave the premises because there is a threat which may materialize shortly. This alert has to be easily recognized by all members of your audience, taking full account of their abilities, wherever they are and whatever they may be doing. The implication here is that you must have the right communication systems in place so that the warning is received and understood without delay or distortion. In most cases, communication means some form of public address system, which may be anything from a simple buzzer or alarm bell up to a full audiovisual set up.

Modern technology provides the opportunity to install what are now known as mass notification systems (MNS) aimed at relatively large scale and mobile audiences, and emergency notification systems (ENS), which are more appropriate for ordinary commercial use. There are a number of such systems on the market, and you would be well advised to do a little research before selecting any of them as the tool for your purposes.

There are also quite sophisticated school bell systems which allow you to play different sounds or tunes according to the message which you want to get across. Another recent development is a whole range of LED messaging systems which can also incorporate loudspeakers to provide combined visual and audio messages.

With such a wide range of communication systems available, it is almost certain that you will find something on the market which can be adapted to meet your needs at a reasonable cost. Here we are using the term "reasonable costs" to indicate the intrinsic value compared to the capital outlay. The message which you have to get across to your sponsors is that this is no time for penny-pinching; people's lives are at stake and emergency communications are a vital element in the battle to preserve their health and safety.

However, you will need tailor your final choice to fit the precise nature of your potential audiences and the environment where they are likely to be, together with the complete range of scenarios, dangers, difficulties, and limitations which they might encounter.

**An alarm should never be used in non-emergency circumstances, and it should certainly not be used for a "surprise test."**

**An Alarm** is the signal which you send when someone has made the decision to invoke the emergency evacuation procedures. It is called an alarm because we expect everybody to react immediately and make for safety, and it should be triggered only when the situation warrants it. Appropriate situations include an emergency, a planned test, or a planned drill. An alarm should never be used in non-emergency circumstances, and it should certainly not be used for a "surprise test."

The use of surprise testing tends to reduce the sense of urgency and dilutes the meaning of the alarm. When the alarm is taken seriously, its use also introduces an element of risk as people scramble for safety, and provides no positive gain over the pre-announced drill or test.

Your challenge is to ensure that everybody will receive, understand, and respond to the signal under all possible circumstances. Such response can be achieved only through the right choice of delivery systems, clear messages, and adequate training for all those involved in the process.

## 6.8.2 Instructions

Once the alarm has been raised, people should begin to evacuate the premises in accordance with the tried and tested evacuation procedures. However, these procedures have been developed in the absence of any specific information regarding the nature, scale or incidental effects of the actual emergency which can affect the procedures. Thus, you may well have a need for supplementary information, additional guidance, and situation updates to be communicated to some or all of the affected population. Therefore, you should be prepared to convey further simple messages as the situation unfolds.

In some situations, it may be advisable or necessary to deliver these subsequent instructions in a restricted or covert manner, targeted at specific groups or areas. Of particular concern is the "lockdown" situation in which you may have to keep people informed of the need to evacuate quickly and quietly in order to avoid exposure to further danger. At the same time, nonessential personnel must be kept out of the area. This requires good communication and coordination between yourself, your organization and the local law enforcement people. Such coordination has to be planned, and preferably rehearsed well in advance.

### Lessons in Emergency Notification from Virginia Tech Tragedy

A tragic chain of events occurred at Virginia Polytechnic Institute and State University in April, 2007. At the time, it was described as the worst ever US shooting incident.

At 7:15 a.m., a gunman, armed with a 9-mm pistol and a .22-caliber handgun, killed a man and a woman in a coeducational residence hall housing 895 people. Some two and a half hours later, police responded to a 911 call reporting that shots had been fired at Norris Hall, an engineering classroom building about a half-mile away on the opposite end of the 2,600-acre campus. The front doors had been chained from the inside, apparently so victims could not escape and police could not enter.

After forcing their way in, police discovered a total of 31 bodies of students and teachers, as well as that of the gunman. A further 15 people were injured.

Throughout this whole episode, confusion reigned, largely due to lack of planning, an air of disbelief, and poor communications. The first e-mail warning was sent out to students, faculty, and staff at 9:26 a.m., more than two hours after the first shooting had taken place at the dormitory. By then, the shooting at the classroom was under way. Unfortunately, the message warned students to be cautious but didn't tell them to stay away from the classroom. The university's president, Charles Steger, said that the administrators and the police initially believed the first shooting was an isolated event and so they didn't foresee the need to close the university. Apparently, they assumed that the gunman had fled the campus. Steger is quoted as saying: "We can only make decisions based on the information you [sic] have at the time. You don't have hours to reflect on it."

Because the building was not equipped with any surveillance equipment, the police were unable to monitor what was happening while the bloody scene was unfolding.

Decisions and actions were based entirely upon false assumptions and a complete absence of any information about what was actually happening. As a result, 33 people died. With more effective communication systems and protocols in place, most of the deaths could have been avoided.

In situations such as this, which are unfortunately becoming all too familiar, it is essential that someone should be able to monitor activities throughout the site or campus. There must also be an adequate communication system to provide up-to-the-minute instructions to the population. These systems must be supported by appropriate reporting and decision protocols in order to ensure that the response is geared in a timely and effective manner to what is actually happening on the ground.

Timely and effective communication across such a large area containing a diverse and mobile population requires a number of alternative, complementary, and supplementary systems. In short, you have to make full use of all the technology which might be available to you and your population within your specific environment. Your communication capability has to be matched to the needs and requirements of those who inhabit, use, or visit your premises. Remember what can happen if there is a delay or breakdown in communication during a violent episode such as happened at Virginia Tech.

An important characteristic of your communication systems in such circumstances has to be flexibility. This flexibility must match the size and diversity of the likely audience. Ideally, you should be able to target specific groups or areas with explicit and unambiguous messages. Such messages must be discreet to avoid causing distress or panic among your people or any intruders who may have invaded your space.

**Legal Aftermath**: In June, 2008, an $11 million settlement by the Commonwealth of Virginia was approved for most of the victims and their families. However, the families of two victims opted out of the settlement, electing to seek damages for negligence in state court. Importantly, on March 29, 2011, the U.S. Department of Education announced the largest possible fine under the federal Clery Act, $55,000, for:

(1) Failure to provide timely warning.

(2) Failure to follow a timely warning policy.

*The fine, the 2008 settlement, and current litigation, all stem from Virginia Tech's lack of a plan to address a reasonably foreseeable emergency that would definitely require an immediate response.*

## 6.8.3 Coded Signals

Earlier, we suggested that in some situations it may be appropriate or necessary to use some form of coded signal. Use of such signals was aimed at those who would be acting as marshals, guides, or rescue workers rather than the workers, residents, or visitors. For the population at large, I believe that coded signals could do more harm than good. While we can expect the emergency response team members to be fully trained in the use of such systems, it is unlikely that everybody else would be familiar with the codes and their meanings. They would be forced to guess at the meaning of a particular signal, or they might choose to ignore it. Either way, the signal will have failed to successfully deliver the right message and could even prompt people to take the wrong action with dire consequences.

The most complex coded signalling that is suitable for the average member of the public is an intermittent sound for a warning or alert, a signal which becomes a continuous sound when the full alarm is sounded and the premises are to be evacuated. Anything more complicated will inevitably cause confusion and disorder. Apart from "standby" and "everybody out," all other emergency messages should use voice or text. Keep the messages brief, repeat them, and stick to plain language.

## Phase 6 Key Actions

- Work with the users to assure the quality and accuracy of the plans before distributing them.
- Develop and deliver an introductory test and exercise program – element testing, end-to-end testing, and a formal exercise.
- Plan how you will deliver and service the various types of plans.
- Make the standard plans or templates available to everyone.
- Customize and deliver personal plans for all those with unique needs.
- Launch the EEP program.
- Train the EEP support team(s).
- Create a permanent program of regularly scheduled full tests and exercises, capturing and applying the lessons learned.
- Make a strong case to your sponsors to adopt the smart fire drill approach.
- Conduct "smart" fire drills, collating the results and reporting your findings.
- Create notification systems to communicate with your audience immediately before, during, and after an emergency.
- Develop and distribute a communications protocol to cover all eventualities, making sure everyone is happy and familiar with the plan.
- Prepare for the long haul, noting what has to be done to maintain the momentum of the EEP program which you have designed and developed to ensure the lives and safety of your population.

# Discussion Questions – Phase 6

This phase is the culmination of the main body of work in connection with EEP; it is also the point at which the style of management will need to change from ambitious and constructive towards more patient and careful. Originally, the aim was to be able to point people in the right direction, but now you need to focus on the detail to ensure that your population has up-to-date and effective instructions on how to reach their emergency destination.

1. During this exercise and maintenance phase, the EEP requires long-term, patient administration rather than short-term, dynamic implementation. Do you see yourself as prepared and experienced enough to deal with the administrative detail? Do you see yourself handing the long-term administration over to someone else? (Bear in mind that if you are in charge of the program, you are still accountable, even if someone else is carrying out the daily administration.) What advice or guidelines would you offer this administrator? What do you consider to be the most important aspects of this role?

2. How do you propose to maintain the momentum over a longer time frame – despite changes in top management? It does happen – even in the best circles. Will a robust policy statement carry the program over from one CEO to the next? Will you have to go back to some of the earlier phases or reiterate their outcomes?

3. What are the immediate lessons to be learned from the Virginia Tech tragedy? List five features that you would recommend be part of any EEP to avoid the recurrence of such an outcome.

4. What are the immediate lessons to be learned from the Freemasons' Hall case study? How would you be able to apply these in a real-life situation? What additional study and practice might be required before you could act confidently on these lessons?

5. If you were speaking to a manager or executive, how would you define a "smart" fire drill as different from the standard fire drill. What are five advantages to the organization that you could give the manager as you argue in favor of a program of "smart" fire drills? Think of your own organization as you consider this question.

# Epilogue

## When I Speak to Your Boss

If ever I got the chance to speak to the chief executive of your organization I would want to convince him or her that Emergency Evacuation Planning is a vital aspect of his or her responsibility for the welfare of their people. There would only be time for me to make five points; I would need to be blunt and get straight to the point.

1. *First, I would want to set the mood for thinking about emergency evacuation issues:*

   "Do you recall the 9/11 collapse of the twin towers and how so many people were unable to reach safety in time? That event, and there are many others, demonstrates the need for effective evacuation procedures."

2. *I might present some scenarios that might trigger memories of recent events for your boss:*

   "It is almost inevitable that there will be an emergency on your premises at some time; it could be a fire, a flood, a riot, a power outage, a storm, an accident, an explosion, or an earthquake. Apart from the moral obligation, there is a legal obligation for you to provide proper emergency evacuation facilities for all those who use or visit your premises. Evidence of adequate preparations and practice will be called for at the inevitable inquiry if anyone comes to any harm."

3. *I would draw your boss's attention to the social media which now provide instant, uncontrollable global coverage of any untoward event:*

"I would be concerned about any media speculation about the cause, attributable to negligence or apathy, of any trauma in relation to failure to evacuate your building successfully, because that could bring your company to its knees. A successful evacuation will probably go unnoticed, but a photograph or tweet of someone suffering will attract worldwide attention."

4. *Executives usually make most of their judgments in relation to value for money or some form of cost-benefit analysis. I would explain*:

   "Emergency Evacuation Planning is a relatively small investment which brings a great deal of benefit and protection. It will enhance employee morale which, in turn, improves customer relationships. It demonstrates a cautious, risk-aware, approach which may be reflected in overheads such as insurance premiums and employee benefits. It is evidence of good corporate citizenship and a caring employer."

5 *The worst-case scenario for an executive is to be confronted by the press without a satisfactory answer to a searching question. My question would be*:

   "How would you cope with journalists who heard that someone had been trapped on your premises? They would assume negligence on your behalf and would be determined to find someone to blame, and the target of the blame is likely to be you!"

# Appendix A

## An Auditing Approach to EEP

Now that you have established your EEP as one of the basic company programs, it will – like every other policy and procedure – need to be subject to a regular audit to evaluate the performance, compliance, and effectiveness of functions, programs, and people. Your EEP audit may take the form of a self-audit procedure or it may be conducted by someone who is from another department or even another company – however, the end goal will be the same: to identify deviations from standard practice or best behavior and to monitor compliance with the relevant rules and regulations. In the world of EEP, such an audit can be constructive and beneficial, ensuring that everything is in place, fully understood, and appreciated by all interested or affected parties. In addition, a regular audit or review will help you to promote the concept of EEP, reinforcing your efforts to embed the subject into the corporate culture.

### The Purpose of the Audit

The purpose of auditing or reviewing the EEP program is threefold:

1. To ensure that there are no serious oversights.
2. To draw attention to the subject and thus make sure it becomes embedded in the organization's culture.
3 To provide evidence of "reasonable precautions" at any subsequent inquiry.

Once an organization has embarked on an EEP program it is less likely that anyone is going to act dishonestly or recklessly, which removes the need for independence, although this may add a certain degree of gravitas to the proceedings. Obviously, the person who carries out the review or audit should not be the same person as the one who has been running the program, simply because of the risk of a blind spot, conflict of interest, or misunderstanding somewhere along the line. A second opinion also makes sure that nothing gets lost in translation. Wherever there is an existing internal audit procedure in place, then you might simply extend that procedure to embrace the EEP program. The auditors will probably want you to provide them with the policy, the work schedule, a list of deliverables together with details of the methodology which is to be applied or followed.

The resulting documentation should provide evidence that a proper process has been established and followed. It will also be helpful to have a series of such documents showing that the regime is regularly monitored. You are looking only to have a report which confirms that an inspection has taken place and the policy and procedures have been followed. If you have conducted exercises or tests, then these should also be recorded in the report. Obviously if your report can offer no evidence of any tests or exercises, then the organization will be seen to be failing in its duty towards those who use or occupy its premises.

Your report should be submitted to the highest authority in the organization because that's where the final responsibility lies and this is the person who could be appearing in court at the end of the day. I also recommend that the report should have appropriately widespread distribution to make sure that everybody is aware of the precautions which are being taken on their behalf.

You should report any shortfalls or errors immediately to whoever has sponsored the EEP program. This report should automatically trigger a coordinated response to deal with the issues. In those rare and unlikely cases in which the sponsor fails to respond proactively, the EEP manager should consider referring the case to a regulatory body. In the UK, for example, this body would either be the industry regulator or the Health and Safety Executive who would take the matter very seriously indeed.

In those places where EEP is a relatively new subject, then I suggest you link up with the existing audit process and ask the internal or external auditors to include EEP within their regular remit in exchange for some form of briefing to enable them to get a grip on the basics of the subject. Where there is no audit procedure at all, then it is less likely that issues such as health and safety will be considered significant.

## Establishing an EEP Audit Routine

With several discrete areas to investigate it should not be difficult to establish a regular audit routine which will ensure you cover every key aspect on a regular basis. I recommend the "little and often" approach in which you review each of these areas, in sequence, covering one area each month. Such an audit requires just a few hours each month and keeps the whole thing alive. Other possibilities include covering two areas on a bi-monthly basis, three areas every quarter, or even six areas every half year.

Since you conduct other audits once a year, you may be tempted to set up yearly audits for EEP as well. However, my experience has been that a full audit organized on an annual basis cannot be relied on to maintain the requisite momentum to ensure that EEP is seen as an integral part of the corporate culture. The only time I would consider an annual audit is where, for some good reason, you wanted to conduct an independent external audit in addition to the regular internal audit or review which you will be conducting.

## The Twelve Stages of EEP Audit Review

To ensure that the overall program is effective, your audit process will need to consider each and every stage, checking that each one makes its proper contribution towards ensuring the long-term safety of your people. I have identified twelve stages in the overall emergency evacuation process which, in your audit, you should evaluate in the same sequence as they are applied – and as they link with one another – in connection with the evacuation and safety of those who may be on or near your premises.

### The Twelve Areas to Investigate in EEP Audit Routine

1. Monitor and Review
2. Awareness and Confidence
3. Exercise and Feedback
4. Alert and Alarm
5. Exit and Escape
6. Rest and Shelter
7. Assembly and Accounting
8. Instruct and Liaise
9. Comfort and Welfare
10. Stand Down and Debrief
11. Record and Report
12. Review and Update

## Stage 1: Monitor and Review

The ongoing routine of an EEP program that is alive and ongoing is monitoring for any events, potential problems, or emergencies which might trigger the need to consider evacuation.

- Identify the person currently responsible and check that he or she is carrying out this work in accordance with the currently published policy and plans.
- Collect tangible evidence that appropriate review and maintenance is ongoing. Monitoring the EEP program includes ensuring that all of the plans, procedures, and associated resources are properly reviewed and maintained on a regular basis.

## Stage 2: Awareness and Confidence

Everybody who uses your premises must be made aware of the EEP program and its implications for them and their friends and colleagues. They must also have confidence that the plans and procedures will ensure that they reach a place of safety without any undue delay or unnecessary risks.

- Check the level of awareness and confidence by speaking to a number of these people. Make sure that you get sample views from members of the various groups within the population. Pay particular attention to the needs of those who might be disabled or disadvantaged, such as the elderly or the very young.
- Check how visitors or strangers are made aware of the evacuation procedures, particularly important if they are likely to be present in large numbers. Does this take the form of passive signage? Audiovisual aids? Explanations from hosts?
- Find out the procedure and determine whether it is being followed and if it is effective. You will need to both observe that the procedure is followed and interview employees and visitors to confirm that the correct messages are being conveyed and understood by the visitors or strangers.

## Stage 3: Exercise and Feedback

It is important that regular exercises are conducted and that there is a feedback process in place which ensures that lessons learned are actually applied. You need to have hard evidence that such a program is in place and that the procedures are being followed. Any oversights in this area may be retrospectively deemed to be criminal negligence if anyone is injured or traumatized in any emergency situation which might arise.

- Check that this aspect of the program is referenced in the EEP policy and properly implemented within the EEP strategy.
- You should be a witness to, and a participant in, some of these exercises.
- Make sure that the feedback procedure is being followed and is effective. Gather evidence that any lessons learned are actually applied.

## Stage 4: Alert and Alarm

The alert procedure is a response to any events which the monitoring suggests have the potential for evacuation. The only way to be certain about the effectiveness and reliability of the alert and alarm procedure is to witness a number of tests or exercises which cover all the options of the various types of event and the range of personnel who may be engaged in this type of activity. Any doubts must lead to further training or modifications to the system.

- Check that the alert procedure enables the person on duty to decide whether or not to put some or all of the population on standby.
- Test the system and its users for the capability to communicate adequate warning or standby messages to all those who may be expected to participate, especially those who will be involved in marshaling or other support activities in the event of an actual evacuation. Make sure that both the system and those who are expected to use it are capable of creating and communicating those messages under emergency conditions. Check that there is an effective stand-down procedure in place which is familiar to all those who might be involved in this stage of the program. Witness this capability before you record it as fit for purpose.
- Ensure that the alarm delivers the correct notice of evacuation to all parties. Observe that it is clearly understood by everyone in the population you are trying to protect.
- Make sure the correct choices of designated exit routes and escape paths are made clear to all the intended recipients. Where applicable, take into account the potential need to be selective about the audience members who are able to receive such alarm messages. In some situations there may be a need for a degree of discretion or secrecy.

## Stage 5: Exit and Escape

The only way you can be sure about the suitability and status of exit routes and escape paths is through regular inspection. Confusion and poor or inadequate communication regarding the choice of exit routes, escape paths, and assembly areas can lead to chaos and possibly disastrous consequences.

- Ensure that all exit routes are marked properly and kept clear for ease of access and use at any time. Alternative routes should be available, and everybody concerned should be familiar with all of them through regular, realistic exercises. Ask to see records of such exercises; if no such records exist, make sure that arrangements are made to conduct such exercises on a regular basis.
- Make sure, through physical inspection, that all exit points are safe to use and that there are alternatives which lead outwards in different directions with escape paths leading to alternative assembly areas.
- Check invacuation areas and rest or shelter points along the route for access and suitable features wherever they are deemed to be an essential part of the evacuation procedure. All those who are likely to be making use of rest points, shelter points, or invacuation areas should be familiar with their use and location through a regular drill or exercise program. These are particularly important where the population includes disabled or disadvantaged people.
- Determine that all escape paths are always open and free from any obstructions or obstacles so that they may be followed easily toward the selected assembly areas.
- Make sure the alarm procedure includes adequate and clear instructions for which paths and assembly areas are to be used in the event of a particular emergency. The communication system must be capable of conveying the correct messages or signals to all interested parties.

## Stage 6: Rest and Shelter

Regularly inspect any rest areas or shelter points along the exit routes or the escape paths wherever these are included as part of the escape procedure. These places are particularly important where the population includes disabled or disadvantaged people.

- Check any invacuation areas for access and suitable features wherever they are deemed to be an essential part of the evacuation procedure. All those who are likely to be using rest areas, shelter points, or invacuation areas should be made familiar with their use and location through a regular drill or exercise program.
- Examine records of such exercises to ascertain that the familiarization is taking place in accordance with the needs of those who will be using these places. If you have any doubts, then you should arrange to participate in some of these exercises and check with the users that they are happy with the exercise program as it is being implemented.

- Carry out regular physical inspections to make sure that these facilities are kept in a suitable condition for use at any time.

## Stage 7: Assembly and Accounting

The long-term safety and welfare of all evacuees depend on the characteristics, maintenance, and accessibility of assembly areas and is of paramount importance in any EEP program. The only way to establish the true facts about the practicality of the assembly and accounting procedures is to witness, or participate in, a realistic evacuation exercise.

- Everybody should be familiar with each of the available assembly areas through regular visits in connection with the evacuation drill or rehearsal program. They should also be able to identify immediately which assembly area is to be used through the messages included in the alarm procedure. To check this, you will need to speak with a representative sample of the population.
- Check whether any other agency or organization might intend to use the identified assembly area for other or similar purposes. If so, make sure that there is capacity to accommodate all those who may intend to gather there. Competition for such spaces may well include the emergency services commandeering the area for such purposes as triage or the parking of emergency support or response vehicles. Talk to the other agency or organization. Where there is doubt, the plan must allow for a constructive alternative.
- Check that there is an effective and reliable accounting procedure in place, along with suitable recording and communication capabilities. In the wake of an emergency, an organization must be able to account for each one of those people who may have been on the premises at the time of the incident. Those in charge must also be able to communicate their findings clearly and promptly to the emergency services. Otherwise, emergency personnel may unnecessarily put their lives at risk in order to establish whether the premises are actually empty.

## Stage 8: Instruct and Liaise

The command and control team needs to be able to communicate with both the evacuees and the emergency services.

- Check that someone is nominated, authorized, and fully prepared to communicate such instructions to the evacuees. Evacuees need instructions about what to expect, what to do, where to go, and when. Also make sure that the facilities are available to enable effective communication with all of the evacuees irrespective of weather or the time of day.

- Make sure there are proper facilities and procedures in place to reassure the evacuees that their friends and colleagues are all safe and accounted for. Get an answer to the question, Does someone make sure that this happens?
- Check that someone is nominated, authorized, and properly informed for a liaison role with emergency services. The emergency services will need to know about the premises, the people, and any hazards they may encounter. Also make sure that emergency services are happy with the way this responsibility is handled.

## Stage 9: Comfort and Welfare

Once the people have reached safety, arrangements need to be made to ensure their comfort and welfare. Check that these important concerns are not left to chance.

- Make sure the assembly area is suitable for use by all of the evacuees, providing adequate shelter and protection from the elements as a minimum. If the population includes those who may be disabled or disadvantaged in some way, they will require somewhere they can sit and rest. Check for availability of toilet facilities and drinking water at or near the assembly area. Ideally there should be some form of catering facilities nearby, especially if the crowd is likely to be there for any length of time.
- Make sure someone is responsible for dealing with the welfare needs of those who have just escaped from a dangerous situation. Someone needs to be prepared to assist with issues resulting from that evacuation. For example, people may be left stranded without vital or important possessions like keys, cash, transportation, and proper clothing. The person empowered to deal with these matters must have the means and authority to act to resolve evacuees' problems.

## Stage 10: Stand Down and Debrief

Whether it is an actual emergency, test, or exercise, someone needs to be properly authorized and organized to inform everybody that the emergency is over, and tell them what is expected of them and what they should expect in return. Your main concern is to ensure that someone is allocated the responsibility for stand-down and debriefing actions and there is somewhere suitable for any necessary meetings to take place.

- Make sure someone has been selected to speak on behalf of the organization in this connection, and that they have the facilities, authority, and capability to carry out this task.

- Make sure the spokesperson is prepared to answer questions about if, when, and where the people should return to their place of work, study, or accommodation. People may need to be advised of changes that have been made or precautions they should take.
- Check that suitable arrangements are in place for an incident debriefing to take place. Everybody should be included or represented at the debriefing so their views can be captured. In those cases where there is a team of marshals, wardens, and helpers they might also take part in separate debriefings.

## Stage 11: Record and Report

Make sure that adequate arrangements are in place to capture and record what happened throughout the emergency (or test or exercise) as well as feedback which was given by participants during the debriefing(s).

- Check that suitable people are charged with responsibility for keeping a history log during the emergency and making notes of what was said at the debriefings. Make sure they have the facilities to create and maintain such records.
- Check that there are procedures and responsibilities in place to ensure all this information is then collated and a formal report is drawn up and distributed to those who need to be advised or apprised of what happened together with any recommendations or lessons which need to be considered or applied.

## Stage 12: Review and Update

The final phase of any emergency evacuation – whether real or as a test or exercise – is the application of the many lessons learned.

- Check that there are adequate reporting and maintenance procedures in place which will ensure that the plans, procedures, and facilities are kept up to date and distributed as expected.
- Check with whoever is responsible for various aspects of this administrative duty, making sure all of them carrying out these tasks without hindrance and with the full support of those they need to work with to ensure that this work is carried out successfully.

# Appendix B

## Standards and Regulations

Over the years, I have sat on a number of standards committees as a subject matter expert and have received training in their methods. In fact, I once qualified as a certified auditor for the British Quality Management Standard BS 5750, which has subsequently been superseded by the international standard ISO 9001.

From my past experience and recent research, I have concluded that standards bodies have tended to avoid applying their relatively rigid disciplined approach to the subject of emergency evacuation planning (EEP). It is a subject which does not lend itself to prescriptive descriptions, definitions, or procedures which can be applied generically. With so many variables in the possible communities, cultures, locations, and their associated populations, it can be really difficult to develop meaningful codes of practice flexible enough to be practical yet sufficiently firm and detailed to be regarded as authoritative. EEP is more of an elusive craft to be molded carefully to fit the place and the people rather than an accurate science to be applied by taking a set of measurements and following a precise formula.

Consequently, most references to EEP in standards are to be found embedded within standards covering other related subject areas. More detailed references to emergency evacuation can be found within regulations and various codes of practice specific to certain industries, types of venue, or particular features, such as elevators, security measures, or communication systems. While a clear link between emergency evacuation and fire regulations exists, fire is only one of many reasons people might need to reach safety in a hurry without exposing themselves to additional dangers.

## Standards Which Refer to EEP

As an example of EEP within another standard, in Europe we find CEN/TS 81-76, which provides safety rules for the construction and installation of lifts (known as elevators in the US) and refers to the evacuation of disabled persons using lifts. This technical specification was produced by the European Committee for Standardization (in French: Comité Européen de Normalisation, CEN). The purpose of this non-profit organization is to foster the European economy in global trading, the welfare of European citizens, and the environment by providing an efficient infrastructure to interested parties for the development, maintenance, and distribution of coherent sets of standards and specifications. Adoption of, and adherence to, common and respected standards encourages the free movement of goods and products, thus stimulating trade in the region. (CEN/TS 81-76 is available from www.cen.eu/cen/Products/Pages/default.aspx)

Another standard from this source is "EN 50172:2004 – Emergency escape lighting systems," which provides detailed guidance on the design, technology, and layout of emergency lighting along exit routes and escape routes. (EN 50172:2004 is available from
(www.en-standard.eu/en-50172-emergency-escape-lighting-systems)

Also based in Europe is the International Organization for Standardization, commonly referred to as ISO. "ISO 23601:2009 Safety identification" is a full international standard establishing design principles for "displayed escape plans" that contain information relevant to fire safety, escape, evacuation, and rescue of the facility's occupants. These plans may also be used, or referred to, by agencies such as the fire and rescue services in case of emergency. (ISO 23601:2009 is available from www.iso.org/iso/home/standards)

These plans are intended to be displayed as signs in public areas and workplaces. Examples of this type of escape plan are included within an annex to the standard. Unsurprisingly, it is pointed out that these examples are for information only and are not to be taken as definitive versions because displayed escape plans need to be customized carefully to suit the location and the population that may be found there.

The US equivalent of ISO 23601 is "ASTM E2238-12 Standard Guide for Evacuation Route Diagrams," published by ASTM International in 2012. Formerly known as the American Society for Testing and Materials (ASTM), this organization is recognized globally as a leader in the development and delivery of international voluntary consensus standards. Today, some 12,000 ASTM standards are used around the world to improve product quality, enhance safety, facilitate market access and trade, and build consumer confidence. (ASTM E2238 – 12 is available from at www.astm.org/Standards/E2238)

Between them, ISO and ASTM provide a wide range of standards recognized and adopted throughout the modern business world. These standards may also be used as the yardstick of good practice and prudent management in any court of law or official enquiry. Even in those jurisdictions which recognize other localized standards, the ISO and ASTM standards would still be regarded as pertinent because all standards bodies are careful to ensure that their various products are compatible and are not at variance with each other.

## BS EN 50172:2004, BS 5266-8:2004

This dual-numbered British and European standard sets out in considerable detail the requirements for emergency lighting in buildings, such as offices and multi-story buildings, open to the public, providing guidance on the illumination of escape routes and safety signs if the normal power supply fails. In addition, the standard specifies the minimum requirements of emergency lighting based on the size, type, and usage of the relevant premises. These requirements also apply to standby lighting, which may be used as emergency escape lighting, and include practical guidance about making lighting more effective in an emergency. (BS EN 50172:2004, BS 5266-8:2004 is available from http://shop.bsigroup.com)

This wide-ranging standard covers a variety of related topics, including emergency escape lighting and the design of emergency lighting, as well as the required system records and log book to provide a credible record which might be called for as evidence. It also gives best practice recommendations on the servicing and testing of emergency lighting systems; this essential aspect of regular maintenance is an area that is often overlooked or left to chance in many standards or codes of practice. Altogether, this standard offers a comprehensive overview of the subject of emergency lighting in public spaces. While this is a European standard, aimed primarily at the European market, its concepts can be applied anywhere in the world to ensure that people in public spaces are able to find their way out to safety through the darkness in an emergency situation.

## EN 54 (European Standard)

"EN 54 Fire Detection and Fire Alarm Systems," a mandatory European standard, specifies in considerable detail the design requirements and required laboratory testing procedures for all components of fire detection and fire alarm systems. Compliance with this standard allows the free movement of a wide range of construction products between countries throughout the European Union market. It was developed and approved by the CEN. (EN54 is available from www.cen.eu/cen/Products).

EN 54 is widely recognized around the world by many countries outside of the European Union, including the Latin American countries, Brazil, most African and Asian countries, and several of the islands in the Pacific Ocean.

"Construction Product Directive 89/106/CE" (a directive from the European Council) mandates that all fire detection and fire alarm equipment be certified against the EN 54 standard by an authorized certification body. Such certification allows the producer to put the CE mark on the product to recognize that the product fulfills all the European Union safety and performance requirements and thus allows the free movement, installation, and usage of such goods throughout the European Economic Area (EEA). (89/106/CE is available from www.europa.eu/legislation_summaries/other/l21184_en.htm)

The CE mark is the only legally recognized condition for trading goods throughout Europe. It aims to ensure the free movement of all construction products within the European Union by harmonizing national laws with respect to the essential requirements applicable to building and construction products in terms of health and safety. The main goal is to ensure that products are introduced to the European market only if they are demonstrably fit for their intended use.

CEN set up a technical committee to define the essential requirements of safety in case of fire. Working with the British Standards Institution (BSI), it produced a series of standards under the generic title "CEN/TC 72 Fire Detection and Fire Alarm Systems." As a result, we now have a complete set of standard requirements and test methods for fire alarm products that have been developed to provide a reliable means of ensuring an acceptable level of safety. This will bring a major benefit to all European countries and is expected to contribute to the reduction of costs related to fires.

The Construction Products Regulation (CPR) has been adopted by the European Commission, replacing the earlier Construction Products Directive (CPD). As a result of the change, the CE mark is due to become mandatory throughout the European Union. After July, 2013, all manufacturers and importers must ensure that their products meet the CE requirements of this new regulation.

The EN 54 standard is published in 27 distinct parts. Those 7 parts relevant to emergency evacuation planning are these:

- EN 54 part 1 Fire detection and fire alarm systems. Introduction.

- EN 54 part 2 Fire detection and fire alarm systems. Control and indicating equipment (Fire alarm control panel).
- EN 54 part 3 Fire detection and fire alarm systems. Fire alarm devices, sounders.
- EN 54 part 14 Fire detection and fire alarm systems. Planning, design, installation, commissioning, use, and maintenance.
- EN 54 part 16 Fire detection and fire alarm systems. Components for fire alarm voice alarm systems. Voice alarm control and indicating equipment.
- EN 54 part 23 Fire detection and fire alarm systems. Fire alarm devices. Visual alarms.
- EN 54 part 24 Fire detection and fire alarm systems. Voice alarms – Loudspeakers.

The other 20 parts of this standard are unrelated to emergency evacuation planning.

## Ongoing Improvements to Emergency Evacuation Procedures (US)

In the US, the National Construction Safety Team Act (H.R. 4687), signed into law October 1, 2002, authorized the National Institute of Standards and Technology (NIST) to establish investigative teams to assess building performance and emergency response and evacuation procedures in the wake of any building failure that has resulted in substantial loss of life or that posed significant potential of substantial loss of life. (Originally set up in 1901, until 1988, NIST was known as the National Bureau of Standards, [NBS].)

H.R. 4687 provides NIST with the responsibilities and authorities modeled on those of the National Transportation Safety Board (NTSB), except that the NIST is set up to deal with the investigation of building failures rather than aircraft accidents. The act gives NIST the responsibility to dispatch teams of experts, where appropriate and practical, within 48 hours after major building disasters and specifically states that at least one member of each team must be a NIST employee. (H.R. 4687 is available from www.nist.gov/public_affairs/factsheet/upload/hr46871.pdf)

The act gives these investigation teams a clear mandate to:

1. Establish the likely technical cause of building failures.
2. Evaluate the technical aspects of procedures used for evacuation and emergency response.

3. Recommend specific changes to building codes, standards, and practices.
4. Recommend any research or other appropriate actions needed to improve the structural safety of buildings and/or changes in emergency response and evacuation procedures.
5. Make final recommendations within 90 days of completing an investigation.

The teams also have the investigative authority needed to access the site of a building disaster, subpoena evidence, access key pieces of evidence such as records and documents, and move and preserve evidence.

However, since NIST is not a regulatory body, it cannot require the adoption of building codes, standards, or practices by state and local governments. Its investigations are not allowed to consider findings of fault, responsibility, or negligence and "No part of any report resulting from such investigation, or from an investigation under the National Construction Safety Team Act, shall be admitted as evidence or used in any suit or action for damages arising out of any matter mentioned in such report." To date, NIST has not elected to exercise its subpoena authority.

## *Two Main Types of Equipment*

We have already mentioned EN 54, a mandatory fire detection and alarm standard in the European Union that is widely used around the world. It covers two main types of equipment, categorized as *initiating devices* and *notification appliances*.

### *Initiating Devices*

The simplest group of initiating devices is made up of those which are manually actuated. Break glass stations, buttons, and manually activated fire alarms are designed and constructed to be located near the exits, where they can be readily identified and operated by whoever is present.

Automatically actuated initiating devices can take many forms, designed and built to respond to any number of detectable physical changes commonly associated with fire: convected thermal energy or movements of unusually hot air; heat detection or high temperatures; products of combustion, such as CO2; smoke detection or particles in the atmosphere; radiant energy; or flame detection. Such devices can also be capable of releasing extinguishing agents such as foam, inert gases, or activating sprinklers. Some automatic devices can be used to monitor and respond to unexpected water-flow. Recent innovations can also make use of cameras and computer algorithms to analyze the visible effects of fire and movement in environments inappropriate for, or hostile to, other detection methods.

### *Notification Appliances*

Normally, evacuation signals are delivered via audible alarm systems producing a distinctive audible tone. Also, they may be equipped with loudspeakers to deliver live or recorded instructions to the occupants and may include a visual signaling capability to produce a warning that can be both heard and seen. This combination of alarm types can communicate the emergency message to everybody, including those who are hard of hearing or have visual difficulties. It is also a safeguard in conditions of poor visibility or excessive noise that make it difficult for occupants to recognize the emergency alarm signal.

In the US, the commonest type of fire alarm or evacuation signaling device uses an audible alarm which produces what is known as a temporal code, i.e. a distinctive sound signal which changes at fixed short intervals. This sound is accompanied by visual notification broadcast throughout the public and common use areas. Obviously, to be effective, all emergency signals need to be quite distinct and easily understood by everyone; otherwise they might be confused with other noises or changes of lighting.

Temporal coding refers to the pattern of sounds or tones that a fire or emergency alarm produces. Normally, the choice of signal pattern is controlled either 1) from a central control panel, or 2) by altering the settings on the device itself by arranging small jumper cables or through dual in-line package (DIP) switches[1] located on the actual signaling mechanism.

Up until 1996, the majority of alarm systems throughout the US produced a steady continuous sound to signal the need to evacuate. At the time, there was no commonly agreed standard or regulation to prescribe any particular type or pattern of sound for audible fire alarm evacuation signals. Although a steady raucous sound was probably the most popular, a number of different structured or coded signaling methods were in regular use as alarm signals.

Special alarm patterns are known as coded signaling and have names which describe, or refer to, their particular structure. Coded signals include *march* time (normally run at 120 pulses per minute or two beats a second). Variants on this theme use 90 pulses a minute or even the more mournful 20 pulses per minute, which gives three seconds on and three seconds off. The pace and rhythm is generally determined by the settings on the control panel. Other common codes include the *hi-lo* code, alternating between two different tones and the slow-whoop,a low tone gradually rising up towards a high tone. A number of variants on these main themes exist, but none is in current use as

---

[1] A DIP switch is a manually operated electric switch that is incorporated, usually on a printed circuit board, as a dual in-line package (DIP). Their advantage is that the settings are easy to change and there are no parts to lose, such as jumper leads. It is also easy to check their settings visually

a fire or evacuation alarm; they have been superseded by a much more distinctive signal, recognized all around the world. Since, international travel has become commonplace, consistent alarms allow people to respond appropriately to a fire alarm, wherever they happen to be at the time.

In December 1987, the ISO approved and published ISO-8201, "Audible Emergency Evacuation Signal". This international standard described an audible signal which, when heard, would clearly mean "evacuate the building immediately." The standard introduced a distinctive pattern of "on" and "off" sounds which could be recognized easily as the signal to leave. (ISO 8201 can be found at www.iso.org/iso/catalogue_detail.htm?csnumber=15293)

An advantage of this standard temporal pattern is that it can be applied to both visual and tactile signals to gain the attention of those experiencing problems with sight or hearing. Various forms of signal can be synchronized, complementing each other so that everyone will realize that it is time to leave the premises without delay.

Subsequently, the essential elements of the ISO standard were adopted by the American National Standards Institute (ANSI)[2] and the National Fire Protection Association (NFPA)[3] when they brought out a joint recommendation in 1996 for a standard evacuation sound pattern, the temporal-three alarm signal or "T-3". Irrespective of the type of sound, this signal consists of three half second pulses followed by a pause of one and a half seconds, repeated for a full three minutes or 180 seconds. The T-3 pattern is also used whenever smoke is detected; smoke detection is often a pre-cursor to fire detection. (Also, the so-called T-4 signal uses a similar pattern of four pulses and is specified for use as an alarm in the event of carbon monoxide detection.) The ANSI and NFPA recommendations for audible emergency evacuation signals are fully described in the standard "ANSI/ASA S3.41 Temporal Pattern" which can be found on the ANSI website at www.webstore.ansi.org. The original ISO standard is available from www.iso.org.

## Voice Alarm Systems

In recent years, as the technology has become more easily available and affordable, voice alarm systems (sometimes referred to as voice evacuation

[2] The American National Standards Institute (ANSI) is a non-profit organization that oversees the development of standards in the United States. They also ensure that U.S. standards are aligned with international standards so that American products can be used worldwide.

[3] The National Fire Protection Association (NFPA) is a United States trade association, with an international membership, that creates and maintains standards and codes for usage and adoption by local governments, fire-fighters and emergency responders.

systems) have become popular in many countries. Most voice alarms are not quite so loud as the traditional horns or bells, although most standards require a certain minimum noise level. An overpoweringly loud voice alarm may be difficult, or even painful, for some people; consequently, they tend to shut out the noise by covering their ears. Voice alarm signals are normally based on one of the standard alarm tones, such as a slow whoop, tone-3, or a chiming bell tone. The choice of signal depends largely on the country and the particular application. The alarm tone is sounded together with a simultaneous voice message announcing that an emergency has been reported and occupants are required to evacuate the building. The message often includes a warning that elevators should not be used in an emergency. Voice evacuation systems can also give specific up-to-date information and/or instructions through a built-in microphone, which gives security personnel a distinct communication advantage over rudimentary horns or bells. This type of alarm system can employ dedicated loudspeakers, perhaps incorporating strobe lighting, or be combined with a regular public address system.

# Modern Trends in Europe

In almost all European countries, voice evacuation systems in public places are required by national regulations. They are mandatory for railway terminals and airports as well as high-rise buildings, schools and colleges, medical facilities such as hospitals, and most other large spaces open to the public. Emergency voice systems were originally developed in Europe during the Second World War II when there was often an urgent need to convey important information and safety messages to large groups of people in emergencies, such as airborne bombing attacks. Indeed, I remember distinctly hearing such broadcasts as a young boy, when my family was ordered to evacuate to a temporary place of comparative safety known as an "air raid shelter."

Voice evacuation systems became common practice during the 1990s, especially for large scale facilities in which public address systems for conveying all sorts of messages were part of the standard operating environment.

Within the UK, voice evacuation systems are governed according to the British Standard (BS) 5839-Part 8, a code of practice for their development and implementation. BS 5839 is available at http://shop.bsigroup.com. Throughout the rest of Europe, the design, operation, and installation of voice evacuation systems is governed by "EN 60849 Sound Systems for Emergency Purposes," developed and published by the European Committee for Electrotechnical Standardization (CENELEC), see www.cenelec.eu. These two standards are similar in content and are further supported by the European harmonized equipment standards family, EN 54 which we referred to earlier. In addition, ISO 7240-16:2007 specifies international requirements for sound system control and equipment in relation to fire protection. This standard is available at www.iso.org/iso/home/store/catalogue_tc/catalogue_detail.htm?csnumber=42978

## *Effectiveness of Voice Alarm Systems*

Between 2005 and 2007, NFPA sponsored a research program focused on developing an understanding of the basic causes of the higher number of deaths seen in high risk groups such as the elderly, those with hearing loss, and those who are intoxicated.

Their findings suggested that a low frequency (520-Hz) square wave output is significantly more effective in arousing high risk individuals. More recent research suggests that strobe lights are not very effective at waking sleeping adults who suffer from hearing loss and further suggests that a different alarm tone is a much more effective means of alerting such people. A number of individuals within the hearing loss community are seeking changes to improve the awakening methods in current use. However, they have not yet managed to come up with a realistic practical alternative to the low frequency square wave signal suggested by NFPA research; so it is likely that the 520-Hz square wave will continue to be the preferred option for some time to come. It is unlikely that a significantly more effective signal will be discovered or developed.

Other modern emergency communication methods include the use of audible textual appliances, employed as part of a fire alarm system that includes emergency voice alarm communications (EVAC) capabilities. These devices combine visual displays of text with verbal messages. High reliability speakers are used to notify occupants of the need for action in connection with a fire or other emergency. These speakers are generally employed in large facilities where general undirected evacuation is considered to be impractical or undesirable. The signals from the speakers are used to direct the occupants' responses. The system may be controlled

[4] UL stands for Underwriters Laboratory which is an independent safety consulting and certification company headquartered in Northbrook, Illinois. Established in 1894, it has participated in the safety analysis of many new technologies, most notably the public adoption of electricity and the drafting of safety standards for electrical devices and components.

UL is one of several companies approved to perform safety testing by the US federal agency Occupational Safety and Health Administration (OSHA). OSHA maintains a l ist of approved testing laboratories, which are known as Nationally Recognized Testing Laboratories.

from one or more locations within the building known as Fire Wardens' Stations, or from a single location designated as the building's Fire Command Center. Speakers are actuated automatically by the fire alarm system in the event of an outbreak of fire; following a pre-alert tone, selected groups of speakers may transmit one or more pre-recorded messages directing the occupants to safety. Where necessary, these messages may be repeated in one or more different languages. Trained personnel activating and speaking into a dedicated microphone can also suppress the replay of automated messages in order to initiate or relay real-time specific voice instructions.

Some fire alarm systems utilize emergency voice alarm communication systems (EVACS) to provide pre-recorded and manual voice messages. Typically, voice alarm systems are used in high-rise buildings, arenas, and other large "defend-in-place" occupancies such as hospitals and detention facilities where total evacuation is difficult to achieve.

Voice-based systems provide emergency response personnel with the ability to conduct an orderly evacuation and notify building occupants of changing circumstances as the situation develops or changes.

In high-rise buildings, different evacuation messages may be played to each floor, depending on the location of the fire. Occupants of the floor where the fire is located, together with those above, may be told to evacuate, while those on lower floors may simply be asked to stand by.

## *Mass Notification Systems/Emergency Communication Systems*

New codes and standards introduced around 2010 (especially the new UL[4] Standard 2572, the US Department of Defense's UFC 4-021-01 Design and O&M Mass Notification Systems, and NFPA 72 2010 edition Chapter 24) have led manufacturers and suppliers of fire alarm systems to expand voice evacuation capabilities to support new requirements for mass notification, including support for multiple types of emergency messaging (e.g., inclement weather emergency, security alerts, amber alerts). (UL Standard 2572 is available from http://ulstandardsinfonet.ul.com/catalog; UFC 4-021-01 is available from www.freestd.us/soft2/792766.htm; NFPA 72 2010 is available from www.nfpa.org/aboutthecodes/AboutTheCodes.asp?DocNum=72)

The major requirement of a mass notification system is to provide prioritized messaging according to the emergency response plan for the local facilities. The emergency response team must define the priority of potential emergency events at a site, and the fire alarm system must be able to support the promotion and demotion of notifications based on this emergency response plan. In the US, emergency communication systems are required to provide visible notification in coordination with any audible notification activities in order to meet requirements of the Americans with Disabilities Act (ADA). (Available from www.ada.gov/pubs/ada.htm.) Recently, many manufacturers have made efforts to certify their equipment to meet these new and emerging standards.

Mass notification system categories include the following:

- Tier 1 Systems are In-Building and provide the highest level of survivability.security alerts, amber alerts
- Tier 2 Systems are Out of the Building and provide the middle level of survivability.
- Tier 3 Systems are "At Your Side" and provide the lowest level of survivability.

Mass notification systems often extend the notification appliances of a standard fire alarm system to include PC-based workstations; text-based digital signage, and a variety of remote notification options including e-mail, text messaging, or interactive voice response messaging. No doubt these systems will continue to evolve as social media continues to develop and expand.

**Emergency elevator service:** Activation of automatic initiating devices associated with elevator operation is normally used to initiate emergency elevator functions, such as the recall of associated elevator cabs or lift cages (as they are known in Europe). Recall will cause the elevator cabs to return to ground level for use by fire service response teams and to ensure that cabs do not return to the floor of the fire incident. Phases of operation include the primary recall (typically to the ground level); an alternate or secondary recall (typically to a floor adjacent to the ground level; illumination of the "fire hat" indicator when an alarm occurs in the elevator hoist-way or associated control room, and in some cases the shunt trip (disconnect) of elevator power (generally used where the control room or hoist way is protected by fire sprinklers). The alternate or secondary recall will be used when the initiation occurred on the primary level.

## UK Fire Alarm System Categories

Many types of fire alarm systems are available, each suited to different building types and applications. A fire alarm system can vary dramatically in both price and complexity. Systems can range from a single panel with a detector and sounder in a small commercial property to an addressable fire alarm system in a multi-occupancy building. Systems have to protect both buildings and occupants.

In the UK, there is a British Standard "BS5839: Fire detection and fire alarm systems for buildings," a code of practice for the design, installation, and maintenance of fire detection and fire alarm systems in buildings. It categorizes the various different types of fire alarm systems commonly available. Category L includes all those designed to protect life, P is used for those to protect buildings, and M is used for manual systems.

**M** Manual systems, e.g., hand bells and gongs. These may be purely manual or manually operated electric devices; the latter may have call points and sounders. They rely on the occupants of the building discovering the fire and acting to warn others by operating the system. Such systems form the basic requirement for places of employment where there is no sleeping risk.

**P1** The system is installed throughout the building – the objective being to call the fire brigade as early as possible to ensure that any damage caused by fire is minimized. Small low-risk areas can be excepted, such as toilets and cupboards less than $1m^2$ ($10.764ft^2$).

**P2** Detection should be provided in all those parts of the building where the risk of ignition is high and/or the contents are particularly valuable. Category P2 systems provide fire detection in specified parts of the building where there is either high risk or where business disruption must be minimized.

**L1** A category L1 system is one designed for the protection of life and has automatic detectors installed throughout all areas of the building (including roof spaces and voids) with the aim of providing the earliest possible warning. A category L1 system is likely to be appropriate for the majority of residential care premises. With category L1 systems, the whole of a building is covered apart from minor exceptions.

**L2** A category L2 system is designed for the protection of life and has automatic detectors installed in escape routes, rooms adjoining escape routes, and high hazard rooms. In medium-sized premises (those sleeping no more than ten residents), a category L2 system is ideal. These fire alarm systems are identical to an L3 system but with additional detection in any area where there is a high chance of ignition, e.g., a kitchen) or where the risk to people is particularly increased (e.g., a sleeping risk).

**L3** This category is designed to give early warning to everyone. Detectors should be placed in all escape routes and all rooms that open onto escape routes. Category 3 systems provide more extensive cover than Category 4. The objective is to warn the occupants of the building early enough to ensure that all are able to exit the building before escape routes become impassable.

**L4** Category 4 systems cover escape routes and circulation areas only. Therefore, detectors will be placed in escape routes, although this may not be suitable depending on the risk assessment or if the size and complexity of a building is increased. Detectors might be sited in other areas of the building, but the main objective is to protect the escape route.

**L5** This is the "all other situations" category. Category 5 systems are the "custom" category and relate to some special requirement that cannot be covered by any other category, e.g., computer rooms, which may be protected with an extinguishing system triggered by automatic detection.

The standard also states that an important consideration when designing fire alarms is that of individual zones. Specifically it describes:

- A single zone should not exceed 2,000m$^2$ (21,528ft$^2$) in floor space.
- Where addressable systems are in place, two faults should not remove protection from an area greater than 10,000m$^2$ (107,640ft$^2$).
- A building may be viewed as a single zone if the floor space is less than 300m$^2$ (3229.2ft$^2$).
- Where the floor space exceeds 300m$^2$ (3229.2ft$^2$) then all zones should be restricted to a single floor level.
- Stairwells, lift shafts, or other vertical shafts (nonstop risers) within a single fire compartment should be considered as one or more separate zones.
- The maximum distance traveled within a zone to locate the fire should not exceed 60m (645.83ft).

## Emergency Egress (US)

Planning for the emergency egress, or evacuation, of people with disabilities must start with awareness. According to the U.S. Census Bureau, about one in five Americans has some kind of disability. This may seem to be a rather high figure until you stop to consider the many definitions of "disability."

Although most of us who are able-bodied tend immediately to think of people in wheelchairs in this context, there are many types and classes of disability.

We must take account of a wide spectrum of disabled persons, including those with temporary, episodic, and chronic disabilities. The temporary category must be familiar to anyone who has ever suffered a sports injury, been pregnant, or suffered from a debilitating infection.

Episodic disabilities include severe allergies or situational disabilities, such as asthma. Situational disabilities may also refer to the effects of an environment in which an otherwise able-bodied person might find himself or herself, such as a high-rise building inducing symptoms of vertigo. Strange or loud noises can also have a negative effect on some individuals, while others can be overwhelmed by obnoxious or powerful smells.

One also has to remember that a significant proportion of the population is not physically capable of walking down many flights of steps.

Though many see the The Americans with Disabilities Act[5] as the legislation that put the rights of people with disabilities on the map, this regulatory process actually began much earlier. In 1947, President Truman established the President's Committee on the Employment of the Handicapped to help veterans with disabilities to move into the workforce. One of the committee's indirect accomplishments was the creation by the American National Standards Institute (ANSI) of the "A117.1, Standard on Accessible and Useable Buildings and Facilities", which can be found at www.iccsafe.org/safety.

The passage of the ADA was followed by the publication of the ADA Accessibility Guidelines (ADAAG), produced and published by the US Architectural and Transportation Barriers Compliance Board, an agency that had been around since 1973. Although it covers areas typically addressed by consensus standards and codes, ADA isn't a code. (ADAAG can be found at http://www.access-board.gov/adaag)

The reason ADA wasn't written like a code was because it was born directly out of the legislative process, not the consensus-based standards-creation process. ADA originally didn't even include provisions for emergency evacuation, though the subsequent ADAAG did.

In the late 1980s, model codes, including "NFPA 101, Life Safety Code" began addressing the issues surrounding accessible egress.

One potentially life-saving aspect of emergency egress for those with disabilities that neither ADA nor the model codes fully address is the use of elevators, the only logical way to get large numbers of people out of a

[5] The Americans with Disabilities Act of 1990

high-rise building quickly.

All major building and safety codes have provisions for accessible elevators. If a building has four or more floors and the fourth and any higher floors are accessible to those with disabilities, there must be an accessible elevator. These types of elevators comprise a vital aspect of accessible means of egress.

NFPA 101 and NFPA 5000 Building Codes both include provisions for emergency evacuation elevators, though currently these provisions apply only to special evacuation needs of air traffic control towers. (NFPA 101 and NFPA are both available from www.nfpa.org/aboutthecodes)

Although NIST has conducted research into egress elevators for about 10 years, no such elevators are currently being manufactured. This could be for several reasons. One of the problems is that the electrical and electronic technologies used to operate and control elevators do not respond well to the effects of water, the commonest and cheapest means of fire protection. There is also the concern that there have been a number of incidents in which firefighters have had close calls or have been badly injured when elevator-shaft doors opened onto an empty shaft or when the elevator brakes failed.

Another difficulty here is that the American Society of Mechanical Engineers' Safety Code for Elevators and Escalators (A 17.1) requires that an elevator's power must be cut automatically and immediately if water is sensed nearby. (A 17.1 can be found at www.asme.org)

The solution to the elevator problem would seem to be simply to use exterior-type elevators, inherently waterproof, but the elevator industry has been reluctant to use such elevators indoors because they're much harder to inspect and maintain properly. Also, the industry is more comfortable with elevators being under the control of trained firefighters during an emergency, rather than untrained civilians.

Another recognized danger is that the emergency power supply for evacuation elevators can't be made utterly foolproof, because power cables could still be affected by the heat of a fire.

A further problem is how one can ensure that an elevator designed and built for use by people with disabilities won't be commandeered by the able-bodied in an emergency.

A challenge for the emergency planner is that almost everything about facility accessibility in the ADA standards for accessible design deals with making sure that people with disabilities can get into and use a building. Almost nothing is said about escape or evacuation from those buildings.

Relatively little of the standards for accessible design ensure that those same individuals can evacuate the building safely in case of a fire or any other type of emergency. It has also been suggested that what little is included is often misunderstood, misapplied, or overlooked, largely because, where such material does appear, it seems to have been added as a casual afterthought rather than a vital aspect of the mainstream thinking.

Three common oversights can significantly affect evacuation for those people who are blind, deaf, hard of hearing, or happen to use wheelchairs.

## Oversight #1: Exit Signs

The ADA standards for accessible design require that all signs designating room numbers, restrooms, and exits should include raised letters and Braille to ensure that people who are blind or partially sighted can find their way around the building. These raised letter and Braille signs are required only if the building already provides signs for these elements. Therefore, if the building does not use and display room names or room numbers, the ADA does not require that they should be added.

If the building does have exit signs, then raised letter and Braille exit signs are required. Most buildings do at least have exit signs because these are required by the relevant fire code. Typically, these would be the normal overhead lighted signs. Since these signs are already present indicating fire exits, then additional signs in raised letters and Braille located next to the exit door jamb and centered 60 inches (152 centimeters) above the floor are required.

Where permanent identification is provided for rooms and spaces, signs shall be installed on the wall adjacent to the latch side of the door. Where there is no wall space to the latch side of the door, including at double leaf doors, signs shall be placed on the nearest adjacent wall. Mounting height shall be 60 in (1525 mm) above the finish floor to the centerline of the sign. The location for such signage shall be so that a person may approach to within 3 in (76 mm) of the signage without encountering protruding objects or standing within the swing of a door.

As a rule of thumb, wherever overhead signs designate the actual exit from the building or floor, raised letter and Braille signs should be installed.

Where overhead signs indicate the path of travel to the exit (that is, where the signs include arrows), raised letter and Braille signs are not required. Research is underway that may find increased value in providing additional ways of finding signage for people who are blind in addition to the signs at the actual exit.

## *Oversight #2: Visual Alarms*

The standards for accessible design require that both audible and visual alarms be included if alarm systems are provided.

If emergency warning systems are provided, then they should include both audible alarms and visual alarms. Sleeping accommodations that comply with the ADA code should have a similar audible and visual alarm system. Emergency warning systems in medical care facilities may be modified to suit standard health care alarm design practice. These systems are to ensure that people who are deaf or hard of hearing are made aware of an emergency in the building.

The most common mode of installation is to include a visual signal with every audible signal. This is certainly a good start, but it does not meet the minimum requirements of the standards for accessible design. Visual alarms are required in restrooms and any other general-usage areas, such as in meeting rooms, in hallways, in lobbies, and any other area for common use.

According to the ADA Standards for Accessible Design, §4.28.1, suitable visual signal appliances shall be provided in buildings and facilities in each of the following areas: restrooms and any other general usage areas such as meeting rooms, hallways, lobbies, and any other area for common use. (ADA Standards are available from www.ada.gov)

Of particular concern are those rooms and spaces that may be occupied and have doors, because the doors can severely limit or prevent the visibility of any alarms located in adjacent hallways. Therefore, a separate visual signal must be located in all such spaces. However, we often find that many commonly used spaces with doors, intended for casual or occasional use, do not have visual alarms, including examination rooms in medical facilities, restrooms, offices, and other such rooms.

## *Oversight #3: Entries & Exits*

The standards for accessible design require that at least 50% of all public building entries, and at least as many exits as are required by the applicable building or fire codes, should be accessible to the disabled.

28 CFR Part 36 – ADA Standards for Accessible Design, §4.1.3(8)

> "In new construction, at a minimum, the requirements in (a) and (b) below shall be satisfied independently:
>
> > (a)(i) At least 50% of all public entrances (excluding those in (b) below) must be accessible. At least one must be a ground floor entrance. Public entrances are any entrances that are not loading or service entrances.

(ii) Accessible entrances must be provided in a number at least equivalent to the number of exits required by the applicable building/fire codes. (This paragraph does not require an increase in the total number of entrances planned for a facility.)"

The standards also require that the same number of accessible means of egress should be provided as are required for exits by local building/life safety regulations.

Where a required exit from a level which may be occupied, but is above or below a level of accessible exit discharge, is not accessible, an area of rescue assistance shall be provided on each such level (in a number equal to that of required accessible exits). A horizontal exit, meeting the requirements of local building/life safety regulations, shall satisfy the requirement for an area of rescue assistance.

People who use wheelchairs must be able to use these entrances or exits in the case of evacuation.

In many cases, buildings have more emergency exits than are actually required by the fire codes. In an evacuation, people who use wheelchairs are not likely to seek out those exits that have been specifically designed with accessibility in mind. We consider that for all practical purposes, every emergency exit in a building should be accessible, regardless of the specific statement of the standards.

A common problem in many facilities is an emergency exit that leads directly out onto a small landing with a step down to a sidewalk or a step-off into a grassy area. Unless this area is designed, and continually maintained, to meet the standards of an area of refuge or area of rescue assistance, people using wheelchairs may be left stranded at the side of the building without any realistic means of getting away from the building unaided.

Clearly, many other elements contribute to the safe and effective evacuation of buildings by people with disabilities. However, we have found that the three elements discussed here are among the most common problem areas. It is important for us all to act as the responsible stewards of accessible evacuation for every person with a disability.

[6] What is described in UK government literature as a "Place of Ultimate Safety" is what we have described throughout this book as an "Assembly Area."

# The Basics of Escaping from Fire (UK specific)

The Regulatory Reform (Fire Safety) Order (RRFSO) 2005 charges the appointed responsible person(s) in control of non-domestic premises with the safety of everyone, whether they are employed in or are visiting the building. Under Article 14 of the RRFSO, this duty of care includes ensuring that "...routes to emergency exits from premises and the exits themselves are kept clear at all times" and that these "emergency routes and exits must lead as directly as possible to a place of safety." In other words, the entire escape route up to, and including, the final exit from a building must remain unobstructed at all times, while the distance people have to go to escape (the travel distance) must be as short as possible. RRFSO is available from www.legislation.gov.uk

Fire exits may or may not be located on the usual routes used when the premises are operating under normal circumstances. The final exit doors should open easily, immediately and, wherever practical, "in the direction of escape," i.e., outwards into a place of safety outside the building. Sliding or revolving doors must not be used for exits specifically intended as fire exits. The emergency routes and fire exits must be well lit and indicated by appropriate signs. In locations that require illumination, emergency lighting of adequate intensity must be provided in case the normal lighting fails and illuminated signs should be used. This is because, as noted in the HM Government publication "Fire Safety Risk Assessment: Offices and Shops" (May 2006): "The primary purpose of emergency escape lighting is to illuminate escape routes but it also illuminates other safety equipment."

# Places of Relative Safety

It is often necessary to devise a temporary place of safety, such as when evacuating high rise buildings. This may be defined as a place of comparative safety and includes any place that puts an effective barrier (normally 30 minutes' fire resistance) between the person escaping and the fire. Examples are as follows:

1. A story exit into a protected stairway or the lobby of approach stairway.
2. A door in a compartment wall or separating wall leading to an alternative exit.
3. A door that leads directly to a protected stair or a final exit via a protected corridor.

A staircase that is enclosed throughout its height by a fire resisting structure and doors can sometimes be considered a place of comparative safety. In these cases, the staircase can be described and known as a "protected route."

However, the degree of protection that enables staircases to be considered a place of comparative safety varies for differing building types, and is normally defined in the relevant codes of practice.

## Place of Ultimate Safety[6]

Ideally, this safe place should be somewhere out in the open air, where unrestricted dispersal away from the building can be achieved. Escape routes should never discharge finally into enclosed areas or yards, unless the dispersal area is large enough to permit all of the occupants to proceed to a safe distance. (Note: a safe distance equates to at least the height of the building, measured along the ground.) Total dispersal in the open air, therefore, constitutes ultimate safety. When inspecting any building, it is important always to follow the escape route to its ultimate place of safety. Plus, the final exits on these escape routes (i.e., fire exits) must have sufficient capacity to ensure the swift and safe evacuation of people from the building in an emergency situation.

## Total Width of Fire Exits

In the UK, two main sources of guidance should be consulted when considering the width of fire exits for your premises: the Building Regulations and the relevant British Standards.

1) Building Regulations adopt an approach based on the maximum number of persons who are likely to be on the premises.

Current building regulations contain guidance on the widths of escape routes and exits for new-build, non-domestic properties and the communal areas in purpose-built blocks of flats in "The Building Regulations 2010, Fire Safety, Approved Document B, Volume 2 – Buildings Other Than Dwelling Houses."

(This document is available from www.legislation.gov.uk/uksi/2010/2214)

The following information is extracted from the above document:

3.18: The width of escape routes and exits depends on the number of persons needing to use them. They should not be less than the dimensions given in the following table:

3.20: Widths of escape routes and exits

In calculating exit capacity, the regulations make the following points:

[6] What is described in UK government literature as a "Place of Ultimate Safety" is what we have described throughout this book as an "Assembly Area."

| Maximum Number of Persons | Minimum Width (in Millimetres) |
| --- | --- |
| 60 | 750 |
| 110 | 850 |
| 220 | 1050 |
| More than 220 | 5 per person |

3.21: If a story or a room has two or more exits it has to be assumed that a fire might prevent the occupants from using one of them. The remaining exit(s) need to be wide enough to allow all the occupants to leave quickly. Therefore, when deciding on the total width of exits needed according to the above table, the largest exit should be discounted.

3.22: The total number of persons which two or more available exits (after discounting) can accommodate is found by adding the maximum number of persons that can be accommodated by each exit width. For example, 3 exits each 850mm (33.46 inches) wide will accommodate 3 x 110 = 330 persons (not the 510 persons accommodated by a single exit 2550mm wide).

2) The British Standards Institution adopts a different approach based on the risk profile of the building.

The BSI "Code of practice for fire safety in the design, management and use of buildings" (BS 9999: 2008) takes a complementary approach to this calculation, based on two main factors: occupancy characteristics and fire growth rate. Combining these two factors creates the risk profile of a specific building, an interpretative approach taking full account of the specific features of an individual building and its usage. This profile is especially significant when you are considering the issue of escape routes and fire exits in existing premises, particularly if they are of an historical or heritage nature. Usually, severe restrictions prevent, or limit, any modifications or alterations to the details of historic or heritage buildings; therefore, the usage or occupancy needs to be made to fit the characteristics of the premises rather than the other way around. (BS 9999 is available from http://shop.bsigroup.com)

The occupancy characteristic is determined principally according to whether the occupants are familiar or unfamiliar with the building (i.e. differentiating between emergency exits and panic exits) and whether they are likely to be awake or asleep.

The standard contains the following table:

**Notes:** *Category 4 is not covered by BS 9999:20008.*

| Occupancy Characteristic | Description | Examples |
|---|---|---|
| A | Occupants who are awake and familiar with the building | Office and industrial premises |
| B | Occupants who are awake and unfamiliar with the building | Shops, exhibitions, museums, leisure centers, other assembly buildings, etc. |
| C | Occupants who are likely to be asleep: | This category is sub-divided as follows: |
| C i | Long-term individual occupancy | Apartments without 24-hour maintenance/management control on site |
| C ii | Long-term managed occupancy | Serviced flats, halls of residence, boarding schools, etc. |
| C iii | Short-term occupancy | Hotels |

**Note:** *Two further categories of occupancy characteristics, "Occupants receiving medical care" and "Occupants in transit" are not covered by BS 9999: 2008.*

The fire growth rate is estimated according to the nature and quantity of combustible materials in a specific building, as follows:

| Category | Fire Growth Rate | Examples |
|---|---|---|
| 1 | Slow | Limited combustible materials |
| 2 | Medium | Stacked cardboard boxes |
| 3 | Fast | Baled clothing, stacked plastic products |
| 4 | Ultra-Fast | Flammable liquids |

Examples of the risk profiles created by combining occupancy characteristic and fire growth rate include the following:

A1: administration office, classroom

A1/A3/A3: storage and warehousing

B1: banking hall, reception area, foyer

B2: theatre/cinema, museum, restaurant

B3: department store, supermarket, furniture store

Cii2: dormitory, study bedroom (e.g., in halls of residence)

The standard notes that the minimum door widths are as given in the table below, with the additional provisos that the total door width should be:

a) not less than the aggregate of the exit widths given in the table.

b) not less than 800mm, regardless of risk profile.

As with the building regulations, the British Standards guidance assumes that, if... then gives the wrong meaning; change to: in a room or story with two or more exits, a fire might prevent the occupants from using one of the exits.

| Risk Profile | Minimum width per person (millimetres) |
|---|---|
| A1 | 3.3 |
| A2 | 3.6 |
| A3 | 4.6 |
| B1 | 3.6 |
| B2 | 4.1 |
| B3 | 6.0 |
| C1 | 3.6 |
| C2 | 4.1 |
| C3 | 6.0 |

Therefore, the remaining exit(s) must be wide enough to allow all of the occupants to leave quickly.

*Example:*

The total number of persons that two or more available exits can accommodate is found by adding the maximum number of persons for each exit width. For example, 3 exits, each 850mm wide, in a building with a B1 risk profile, would accommodate 472 persons, as illustrated by the following calculation:

- 850/3.6 = 236
- discount one exit
- 2 x 236 = 472 (not the 708 who could be accommodated through a single exit 2550mm [i.e. 3 x 850mm] wide in a building with a risk profile of B1)

You will notice that this example suggests a larger maximum number of persons (236) can be accommodated by an exit width of 850mm in a building with a low risk profile than as stated in the building regulations, which only allows for 110 persons using an exit of this width. This suggests that the building regulations' estimate is based on the worst case scenario from the point of view of fire growth rate.

## Means of Escape for Disabled People in Public Places

In public places, The Regulatory Reform (Fire Safety) Order 2005 requires that a specific responsible person should be appointed and when that person is conducting a fire risk assessment and considering the means of escape from fire, the assessment should incorporate the recommendations of:

- "BS 8300:2009. Design of buildings and their approaches to meet the needs of disabled people" from the British Standards Institution.
- The guide "Means of Escape for Disabled People," which can be downloaded from the Department of Communities and Local Government web site.(Go to www.gov.uk/government/organisations/department-for-communities-and-local-government

These are not statutory documents, but they do provide authoritative guidance on the design and management of buildings to enable the safe evacuation of people with disabilities. This guidance includes coverage for people with hearing and sight loss. It also includes application to existing buildings. Copies of the British Standard may be obtained on loan from local libraries within the UK, or purchased from the British Standards Institution. (at http://shop.bsigroup.com)

The following guidance should be read in conjunction with BS 8300:2009, from the above mentioned guide, "Means of Escape for Disabled People." The guide includes the recommendation that evacuation plans should be devised only by persons who are familiar with the location and the people involved.

- Disabled people, like everyone else, should always have available a safe means of escape in the event of fire.
- The nominated person in charge must, with the assistance of the appointed responsible person, make the best practicable arrangements for ascertaining what areas are likely to be used by disabled people. Furthermore, in consultation with these people, the nominated person is responsible for making adequate arrangements for their evacuation in the event of fire. These arrangements must be tested.
- A personal fire evacuation plan (PEEP) should be drawn up for every disabled person or group of disabled people in the building. Regular building users who are disabled should receive their own copy of a PEEP. If the building is one with a large number of visitors, then simple relevant fire evacuation instructions should, so far as possible, be handed to disabled visitors by reception staff on arrival.
- So far as is reasonably practicable, fire compartmentalization in buildings used by disabled people, and any other arrangements, must comply with "BS 8300:2009 Design of buildings and their approaches to meet the needs of disabled people."
- Lifts (elevators) must not be used in the event of fire unless they meet the special requirements of "PD 7974-6:2004 – The application of fire safety engineering principles to fire safety design of buildings. Human factors. Life safety strategies. Occupant evacuation, behavior and condition." (PD 7974-6 is available from http://shop.bsigroup.com)
- A sufficient number of people should be trained in advance in giving assistance to disabled people so that the necessary number would always be present in the event of an emergency.
- Where necessary, arrangements must be made for the presence of the disabled person(s) to be known to those who would give assistance. This could be done with an in-out tally at the entrance or by informing someone, providing that the desk or office involved is permanently manned during the day. In some cases, for

example ensuring that deaf or blind people are helped out, a floor warden system may be more appropriate.

- The placing of restrictions on disabled people, requiring them to be accompanied at all times by potential helpers, should where possible be avoided. In buildings with good fire compartmentalization, i.e., separation of fire compartments, it will usually be possible for people to work unaccompanied, provided that there are adequate numbers of potential helpers elsewhere within the building. However, disabled people who would need assistance to leave in an emergency should not use buildings at times when insufficient helpers may be present to assist their evacuation (e.g., evenings and weekends). Also, if compartmentalization in one area does not reach the standard of BS 8300:2009, then it may be necessary to require that a disabled person uses the area only when sufficient numbers are immediately at hand. A disabled person might use a particular floor for normal work and need to visit other places during the course of his or her work.
- Disabled people should not use any part of a building where it would be difficult for them, even with help, to escape in the event of fire. The use of basements by wheelchair users, where there is no basement level exit, is likely to be an example of this. Activities assigned to such areas should be moved to different areas, so far as reasonably practicable, in order to avoid excluding disabled people from those activities.
- In the case of work above ground floor level by people who use a wheelchair or have difficulty with stairs, arrangements should be based on horizontal movement away from fire through fire-resisting doors to an area of refuge. BS 9999 (previously referred to indicates the layout requirements for this. Emergency evacuation procedures ought to be based on the following principles:
  - When the fire bell rings the disabled person requests assistance from anyone nearby to help in evacuation. The disabled person and helpers wait, without causing obstruction, in a place near the stairs, until the other occupants have gone down and the disabled person is then carried or helped downstairs. It may be necessary to provide one or more special evacuation chairs for this purpose.
  - If insufficient helpers are on hand the disabled person should move to the main stairwell, unless there are signs of smoke or

fire, in which case the stairwell furthest away from the fire is to be used. Another stairwell might be used, if this had been considered by prior agreement with the emergency party to be more convenient. The disabled person should remain in the stairwell and wait for assistance.

- The emergency party gathers, and if the disabled person is known to be in the building, the party should go to the pre-arranged staircase or, if that is in or very near the fire, to the alternative staircase and carry the disabled person down.
- A fire safety adviser can help in the application of this code to particular circumstances, and should be consulted in any case where it appears that building modifications might be required to provide a safe means of escape for disabled persons.
- Disabled people should be taken to include all those who are temporarily disabled through injury or illness.

## Legal Overview

The UK Fire and Rescue Service's role as an enforcing authority is to ensure that the means of escape in case of fire, and the associated fire safety measures provided for all the people who may be in a building, are both adequate and reasonable, taking into account the circumstances of each particular case.

Under current fire safety legislation it is the appointed responsible person, as defined by the Regulatory Reform (Fire Safety) Order 2005, who must arrange a fire safety risk assessment; provide an emergency evacuation plan for all people likely to be in the premises, including disabled people; and determine how that plan will be implemented. Such an evacuation plan should not rely, or be dependent, upon the intervention of the Fire and Rescue Service to make it work. In the case of multi-occupancy buildings, responsibility may rest with a number of responsible persons for each occupying organization and with the owners of the building. It is important that they cooperate and coordinate evacuation plans with each other. This could present a particular problem in multi-occupancy buildings where the different escape plans and strategies need to be coordinated from a central point.

**The UK Disability Discrimination Act 1995/2002** (**DDA**) underpins the above fire safety legislation in England and Wales by requiring that employers, or organizations that provide services to the public, take full responsibility for ensuring that all people, including disabled people, can leave the building they control safely in the event of a fire. Where an employer or a service provider does not make provision for the safe evacuation of disabled people from its premises, this may be viewed as discrimination. It may also constitute a failure to comply with the requirements of the fire safety legislation mentioned above.

Public bodies have an additional duty, called the Disability Equality Duty (DED), which, since 2006, requires them to proactively promote the equality of disabled people. This requires them to ensure that disabled people do not face discrimination by not being provided with a safe evacuation plan from a building.

## Fire Engineering Principles

PD 7974-6:2004 – "The application of fire safety engineering principles to fire safety design of buildings" is a published document developed and published by the British Standards Institution. As a published document it is not prescriptive; it is intended to provide guidance to designers, regulators, and fire safety professionals on the engineering methods available for the evaluation of life safety aspects of a fire safety engineering design in relation to evacuation strategies. It describes good practice rather than states obligations. (For more details on DDA go to https://www.gov.uk/definition-of-disability-under-equality-act-2010).

Should a fire occur in which occupants of a building might be exposed to fire effluent and/or heat, the primary objective of the fire safety engineering strategy is to ensure that such exposure does not significantly impede or prevent the safe escape (if required) of essentially all of the occupants, without their experiencing or developing serious health effects.

Advice is presented on the evaluation and management of occupant behaviour, particularly escape behaviour, during a fire emergency and for the evaluation of occupant condition, particularly in relation to exposure to fire effluent and heat.

This published document addresses the parameters that underlie the basic principles of designing for life safety and provides guidance on the processes, assessments and calculations that are necessary to determine the location and condition of the occupants of the building, with respect to time. It also provides a framework for reviewing the suitability of an engineering method for assessing the life safety potential of a building for its occupants.

Other parts of this publication cover further related aspects of fire engineering. Sub-system one (PD 7974-1) provides information about estimating the rate of production of heat and combustion products from the fire source. The aim of Sub-system two is to provide design approaches to enable the estimation of the spread of the combustion gases within, and beyond, the room of origin and to evaluate their properties, i.e. temperature, visibility, and concentration of toxic products. Such information can be used to calculate the time elapsed from the detection of a fire to conditions developing which would be dangerous to occupants of the building in question. This will enable the design of fire safety measures to ensure that sufficient time is available for escape. It also provides information that will allow property issues to be assessed.

## Disabled Access in the UK

"BS 8300: 2009+A1:2010 – Design of buildings and their approaches to meet the needs of disabled people" is a code of practice dealing with the design of buildings and their ability to meet the requirements of disabled people. By offering best-practice recommendations, this standard explains how architectural design and the built environment can help disabled people to make the most of their surroundings. It looks at how some facilities, such as corridors, car parks and entrances, can be designed to provide aids for the disabled. It also demonstrates how additional features, including ramps, signs, lifts and guard rails, can be installed in order to improve the environment for the benefit of the disabled. (See previous reference to BS8300 for URL details.)

The comprehensive requirements set out within BS 8300, take account of a wide range of different types of disabilities and cover the use of public buildings by disabled people who are residents, visitors, spectators or employees. This standard's recommendations include such aspects as parking areas, setting-down points and garaging, access routes to and around all buildings, as well as entrances and interiors. BS 8300 gives full consideration to all the features and qualities needed to enable or assist disabled people whenever they are approaching or leaving such a building. It also covers the relevant routes to all the facilities that are associated with these buildings.

Although the recommendations of BS 8300 are specifically aimed at public buildings, the same concepts can be applied to any private building. Although this may not be a legal requirement it would prove beneficial to any disabled people who might have access to such a building.

BS 9999:2008 is the British fire safety code of practice for building design, management and use. This standard outlines ways of meeting the fire safety legislation through adopting a more flexible and thoughtful approach to the design, management, and general use of a building. BS 9999:2008 provides a risk-based structure that takes various human factors into account, including improving emergency exit access for use by disabled people. The recommendations of this standard can be adopted and used in and around existing buildings, or they can be implemented during the design stage for new buildings or extensions. They can also be applied when making alterations, extensions and changes of use of an existing building. In addition this standard provides a useful assessment tool to ensure that the fire safety strategy remains robust throughout the many minor and occasional major changes which may occur in the management or usage of a building over time. (See previous reference to BS 9999 for URL details)

BS 9999:2008 is based on UK government guidance and provides a best practice framework for fire safety. The standard outlines ways of testing all aspects of your fire strategy, including easy access to exits, to ensure the safety of people who may be in and around the buildings. This includes guidance on how to manage fire safety throughout the entire lifecycle of the building – starting with the original design or redesign, through to fire system assessment and maintaining an effective fire detection system. BS 9999:2008 also gives useful guidance on the training of employees in fire safety, organizing an efficient evacuation plan, and allocating leadership responsibilities. The recommendations and guidance given in this British Standard are intended to safeguard the lives of both the building occupants and the firefighters who may need to respond to an emergency in or near the building.

At 458 pages in length, this standard is comprehensive and goes into a great deal of useful and practical detail. Although it is specifically aimed at meeting the expectations of UK government legislation and guidance, its principles and concepts regarding the ongoing management, usage, and training in connection with fire safety could be sensibly applied almost anywhere.

## ASTM E2513 - 07(2012) Standard Specification for Multi-Story Building External Evacuation Platform Rescue Systems

This US standard covers the specifications, safety requirements, performance, design, practices, marking instructions, and test methods for multi-story building external evacuation platform rescue systems (PRS) for the emergency escape of persons who cannot use the normal means of egress to a safe area and for the vertical transport of emergency responders. This standard is applicable only to those PRSs which are permanently installed, designed for multi-cycle and repetitive use, and where the descent is controlled to limit the speed before arrival at a floor or landing zone. Conversely, this specification does not cover platform devices that are used primarily for purposes other than emergency evacuation and/or access, such as helicopters or other flying platforms, platforms that can be transported to or between buildings during operations, or a PRS which uses driving methods other than a positive drive such as drum and ropes. (ASTM E2513 is available at www.astm.org/Standards/E2513)

There is also an US standard for what are known as controlled descent device systems (CDD). This is known as "ASTM E2484 Standard Specification for Multi-Story Building External Evacuation Controlled Descent Devices." (Available from www.astm.org/Standards/E2513)

This standard defines the material requirements, performance, design, marking instructions, test methods, ancillary components, requirements for the installation, periodic maintenance when installed, and instructions for the use of multi-story building external evacuation controlled descent device (CDD) systems for the emergency escape of persons who cannot use the standard exit facilities in multi-story buildings. This standard does not apply to personal escape parachutes, ropes, chain ladders, or rappelling devices. Neither does it apply to ancillary components which may be used with, or included in, CDD systems, such as harnesses, connecting hardware, signage, special evacuation openings, personal protection equipment, or other devices and components used in conjunction with CDD systems.

While these external evacuation and rescue systems are still comparatively rare, it is reassuring to know that formal standards exist for what could be some fairly hazardous contraptions if there were no rules or guidelines in place. No doubt we shall see an increase in these sorts of device as buildings of the future become ever higher and more complex in their design

Other relevant standards include:

- "ASTM F1297 Standard Guide for Location and Instruction Symbols for Evacuation and Lifesaving Equipment." This US standard describes the proper use of symbols that should be used to identify the location and operation of lifesaving equipment particularly related to the evacuation of personnel in a marine environment. (This standard is available from www.astm.org/Standards/F1297)

  It describes, with illustrations, the full range of symbols to be used whenever graphic representation could assist personnel in locating their emergency stations and equipment and in the operation of such equipment. It also describes how and where these symbols should be placed in conspicuous locations in the vicinity of survival craft, their launching controls, or other lifesaving equipment.

- An international standard, "ISO 3864-1:2011 Graphical symbols," describes the design principles for safety signs and safety markings. This standard is applicable to signage in all types of locations. The reasoning behind this universal approach is that the same or similar signage should be used in all situations so that people will become familiar with it through constant and consistent exposure, thus avoiding confusion when in a strange or unfamiliar location. It also specifies the colors and symbols to be used for safety signs and safety markings.

  This ISO standard provides a full description regarding the colorimetric and photometric properties of the basic materials to be used in the design and manufacture of safety signs.

- In addition, there is a British Standard "BS 5499-4:2000 Safety signs, including fire safety signs. Code of practice for escape route signing." This is a part of the BS 5499 series, and provides recommendations and guidance on the selection and use of escape route signs conforming to BS 5499-1. Its recommendations are designed to satisfy the requirements of the Health and Safety (Safety Signs and Signals) Regulations 1996 and existing fire safety legislation in the United Kingdom. It gives advice on the selection and use of the appropriate graphical symbol, the use of supplementary text to assist in the interpretation of signs, and the use of arrows to provide additional directional information. (BS 5499 is available from shop.bsigroup.com; the Health and SafetRegulations are available from www.hse.gov.uk/involvement/1996.htm)

- It does not, however, stipulate if and when escape route signs will be required.

  This standard applies to all those premises where the formal risk assessment, carried out under the management of health and safety, and in particular the Fire Precautions (Workplace) Regulations, has indicated a need for escape route signs to form an integral part of normal working procedures. It is intended to cover the use of escape route signing systems within such premises.

  It provides recommendations for the selection of the appropriate type of sign, the location of those signs, their mounting positions, lighting, and maintenance. The recommendations within this standard also cover the use of high-mounted signs, but it does not cover determining whether there is a need for escape route signing.

### *Australian Standards and Guidelines*

Over the years, I have collaborated regularly with a number of business continuity, risk management, and emergency planning professionals from Australasia, and I have often been impressed by the very pragmatic and user-friendly standards and regulations in that part of the world. Their main reference standard relating to our field of interest is "Australian Standard AS 3745 – 2010: Planning for Emergencies in Facilities." (This standard is available from www.standards.org.au)

This standard, prepared by a panel of experts in the field of emergency response, sets out the role of the emergency planning committee and the response procedures of the requisite emergency control organization.

## Basic Requirements of AS 3745

The standard sets out guidelines for:

1. Appointing an emergency planning committee.
2. Establishing an emergency control organisation.
3. Preparing emergency plans and procedures.
4. Establishing roles for key personnel.
5. Establishing education and training requirements.

AS 3745 is intended to apply to all types of buildings including complex multi-tenanted facilities. However, it is anticipated that the majority of plans required to be approved by local government will be for relatively small self-contained single-tenant premises.

In approving emergency plans, local government should identify specifically that items 2 to 5 above have been addressed and that the plan is generic and can be applied to all emergencies. In practice, it is not uncommon for plans to only address "fire" situations and thus overlook all the other types of incident which may necessitate the invocation of emergency plans and procedures.

AS 3745 is complemented and supported by a freely available series of emergency manuals, published by the Attorney General's Department. Manual number 11 in this series specifically covers the subject of evacuation planning. The foreword includes the following sound advice:

> "As more and more people are becoming affected by the impact of emergencies and disasters across the globe, it is increasingly imperative that response and recovery agencies, organizations and individuals in the community focus on preparedness for a

> wide range of situations. Evacuation is a significant element of this focus.
>
> Effective planning is integral to building the resilience of organizations and communities through their active involvement in the process. In the event of a hazard impact or threat, the evacuation process is vital to saving lives and preventing injury. As part of a risk management strategy, evacuation planning can be used to mitigate the effects of an emergency or disaster on communities."

The "Australian Emergency Manual #11 – Evacuation Planning" includes comprehensive guidelines to assist in the formulation of evacuation plans at all levels. These guidelines are designed to assist not only the emergency services, but local government, state and territory government agencies and Australian government departments in developing new, and revising existing, evacuation plans.

## Summary

At first glance it seems as though few standards appear to deal specifically with our subject of emergency evacuation planning. However, when one looks more carefully at the content of the wide range of related standards out there, it is not difficult to find guidance and recommendations to which you can refer as authoritative sources when justifying or explaining what you are doing.

We have provided you with an overview of the pertinent standards and regulations, but you should also be aware that this is a fertile and often vibrant area with new material always on the horizon.

## Standard References

The standards, regulations and guidance mentioned in this chapter can be accessed via the following URLs which are listed in alphanumeric order.

"Americans with Disabilities Act of 1990" can be found at: - www.ada.gov/pubs/ada.htm

"ADA Accessibility Guidelines for Buildings and Facilities (ADAAG)" can be found at: - www.access-board.gov/adaag

"ADA Standards for Accessible Design, §4.28.1" can be found at: - www.ada.gov/adastd94.pdf

"ANSI/ASA S3.41 Temporal Pattern" can be found at: - www.webstore.ansi.org.

"ASME Safety Code for Elevators and Escalators" can be found at: - www.asme.org

"ASTM E2238 - 12 Standard Guide for Evacuation Route Diagrams" can be found at: - www.astm.org/Standards/E2238

"ASTM E2484 Standard Specification for Multi-Story Building External Evacuation Controlled Descent Devices" can be found at: - www.astm.org/Standards/E2513

"ASTM E2513 - 07(2012) Standard Specification for Multi-Story Building External Evacuation Platform Rescue Systems" can be found at: - www.astm.org/Standards/E2513

"ASTM F1297 Standard Guide for Location and Instruction Symbols for Evacuation and Lifesaving Equipment" can be found at: - www.astm.org/Standards/F1297

"AS 3745 – 2010: Planning for Emergencies in Facilities" can be found at: - www.standards.org.au/search/Results.aspx?k=3745

"Australian Emergency Manual #11 - Evacuation Planning" can be found at: - www.em.gov.au/Documents/Manual11-EvacuationPlanning.pdf

"A117.1, Standard on Accessible and Useable Buildings and Facilities" can be found at: - www.iccsafe.org/safety

"BS EN 50172:2004, BS 5266-8:2004 Emergency escape lighting systems" can be found at:
http://shop.bsigroup.com/ProductDetail/?pid=000000000030154613

"BS 5499-4:2000 Safety Signs, Including Fire Safety Signs. Code of Practice for Escape Route Signing" can be found at: -
http://shop.bsigroup.com/en/ProductDetail/?pid=000000000030054548

"BS 5839-9:2011 Fire detection and fire alarm systems for buildings" can be found at: -
http://shop.bsigroup.com/ProductDetail/?pid=000000000030136499

"BS 8300:2009. Design of Buildings and Their Approaches to Meet the Needs of Disabled People" can be found at: -
http://shop.bsigroup.com/en/ProductDetail/?pid=000000000030217421

"BS 9999: Code of Practice for Fire Safety in the Design, Management and Use of Buildings" can be found at: -
http://shop.bsigroup.com/ProductDetail/?pid=000000000030158436

"Building Regulations 2010" can be found at: -
www.legislation.gov.uk/uksi/2010/2214

"CEN/TS 81-76 Safety rules for the construction and installation of lifts" can be found at: - www.cen.eu/cen/Products/Pages/default.aspx

"Construction Product Directive 89/106/CE" can be found at: - www.europa.eu/legislation_summaries/other/l21184_en.htm

"DDA - The UK Disability Discrimination Act 1995/2002" can be found at: - www.legislation.gov.uk/ukpga/2010/15/section/6

"EN 50172:2004 _ Emergency escape lighting systems" can be found at: - www.en-standard.eu/en-50172-emergency-escape-lighting-systems

"EN 54 Fire Detection and Fire Alarm Systems" can be found at: - www.cen.eu/cen/Products

"EN 60849 Sound Systems for Emergency Purposes" can be found at: - www.cenelec.eu

"Fire Safety Risk Assessment: Offices and Shops" can be found at: - www.gov.uk/government

"Health and Safety (Safety Signs and Signals) Regulations 1996" can be found at: - www.hse.gov.uk/involvement/1996.htm

"H. R. 4687, The National Construction Safety Team Act," can be found at: - www.nist.gov/public_affairs/factsheet/upload/hr46871.pdf

"ISO 23601:2009 Safety identification" can be found at: - www.iso.org/iso/home/standards

"ISO 3864-1:2011 Graphical symbols" can be found at: - www.iso.org/iso/catalogue_detail.htm?csnumber=51021

"ISO 7240-16:2007 Fire detection and alarm systems" can be found at: - www.iso.org/iso/home/standards

"ISO-8201 Audible Emergency Evacuation Signals" can be found at: - www.iso.org/iso/home/standards

"Means of Escape for Disabled People" can be found at: - https://www.gov.uk/government/uploads/system/uploads/attachment_data/file/14898/fsra-escape-disabled.pdf

"NFPA 101, Life Safety Code" can be found at: - www.nfpa.org/aboutthecodes/AboutTheCodes.asp?DocNum=101

"NFPA 5000: Building Construction and Safety Code" can be found at: - www.nfpa.org/aboutthecodes/AboutTheCodes.asp?DocNum=5000

"NFPA 72: National Fire Alarm and Signaling Code" can be found at: - www.nfpa.org/aboutthecodes/AboutTheCodes.asp?DocNum=72

"PD 7974-6:2004 _ The Application of Fire Safety Engineering Principles to Fire Safety Design of Buildings" can be found at: -
http://shop.bsigroup.com/en/ProductDetail/?pid=000000000030041515

"Regulatory Reform (Fire Safety) Order 2005" can be found at: -
www.legislation.gov.uk/uksi/2005/1541

"UFC 4-021-01: Mass Notification Systems" can be found at: -
www.freestd.us/soft2/792766.htm

"UL 2572: Mass Notification Systems" can be found at: -
http://ulstandardsinfonet.ul.com/scopes/scopes.asp?fn=2572.html

# Glossary

**alert** A formal notification that an incident has occurred that may develop into a disaster or a crisis.

**assembly area** A nearby safe space, designated as a gathering point for personnel.

**assembly area requirements (AAR)** The required characteristics of suitable assembly areas where people may be expected to gather in safety.

**assembly time objective (ATO)** The amount of time allowed for people to reach their assembly area safely.

**back-up** Alternative option, measures, or resources held in reserve for emergency purposes.

**building denial** Any condition which causes denial of access to the building or the working area within the building.

**business continuity** The capability to continue essential business functions under all circumstances.

**business continuity management (BCM)** Those disciplines, processes, and techniques which seek to provide the means for continuous operation of the essential business functions under all circumstances.

**business continuity plan (BCP)** A comprehensive plan designed to ensure continuous operation of the essential business functions. The BCP is normally a high-level plan that co-ordinates and controls the responses of individuals and teams to a disastrous event.

**business impact analysis (BIA)** A process for evaluating the costs of a disaster or an emergency. The results may be expressed in financial and non-financial terms.

**contingency plan** Any plan of action to be followed in the event of a disaster or emergency. (A generalization, a non-specific term).

**crisis** An abnormal situation or a perception that poses a threat to the operations, staff, customers, brand, image, or reputation of an enterprise.

**crisis management plan** A plan of action designed to support the crisis management team when the members are dealing with a crisis, or a potential crisis.

**crisis management team (CMT)** A group of executives who assume responsibility for the long-term survival and the image of the enterprise.

**decision point** A point along an exit route or an escape path at which people may have to make a choice of direction or how to proceed.

**declaration (of an emergency)** A formal statement that a state of emergency exists.

**disaster** Any accidental, natural, or man-made malicious event that endangers or disrupts normal operations, such as to threaten failure of the enterprise.

**disaster recovery (DR)** The process of attempting to return critical functions of the business to a state of normal, or near normal, operations.

**disaster recovery plan (DRP)** A plan to support the resumption, or recovery, of a specific operation, function or process of an enterprise.

**egress** A term sometimes used to describe the exit or the way out of a site, a building, or an area within a building.

**emergency** A situation in which there is a threat of destruction or damage to persons or property. The state of emergency continues until there are no further primary or secondary threats.

**emergency assembly area** A nearby safe space, designated as a gathering point for personnel in the event of an emergency.

**emergency control center (ECC)** The location from which an emergency is managed and controlled; it may also serve as a reporting point for deliveries, services, press, visitors, and all external contacts.

**emergency evacuation planning (EEP)** The process for producing an emergency evacuation plan, also known as an emergency egress plan. Sometimes the acronym is used to signify the product, i.e., the plan itself, rather than the process.

**emergency impact analysis (EIA)** The process of prioritizing potential threats to the physical environment.

**emergency management plan (EMP)** A plan that supports the emergency management team by providing the members with information and guidelines.

**emergency management team (EMT)** The group of management staff that commands and manages the resources needed to deal with an emergency situation.

**emergency response plan (ERP)** An operational level plan that provides instructions and information regarding the procedures to be followed in the event of an emergency.

**emergency services** In the context of this book, the term is used to describe all groups that are trained and responsible for responding to an emergency situation, including such groups as the police, fire and rescue, ambulance, paramedics, first aid teams, and coast guard. In a large-scale, or environmental, disaster the military might also be included within, or co-opted into, this group.

**emvacuation** A made-up word used to describe the whole subject of emergency evacuation planning. It covers exit routes, escape routes, invacuation, internal refuges, and assembly points.

**escape requirements analysis (ERA)** The process of estimating the overall requirements for a successful evacuation.

**escape route** A preferred route or pathway that leads from an exit point to an assembly area. People use escape routes to reach safety once they are out of the building or workplace.

**evacuation** The process and procedures that are used in an emergency to alert people and get them from their normal place of work to a place of safety.

**evacuation plan** A description of the procedures and processes that are used to alert people and get them from their normal place of work to a place of safety.

**exercise** A procedure, routine, or drill which is carried out for training, learning, and improvement.

**exit route** A preferred route or pathway that leads from the normal place of work to an emergency exit point. People use exit routes to get out of the building or workplace.

**fire exposure analysis (FEA)** A process which can be used to demonstrate the value of effective emergency evacuation procedures.

**imminent catastrophic event (ICE)** A term used to describe the situation where there is some warning of a significant threat to a building, usually in relation to a non-fire type of threat.

**impact** Cost to the enterprise, not necessarily measured in purely financial terms.

**incident** Any event that is, or is likely to become, an emergency or a disaster.

**ingress** A term sometimes used to describe the entrance or a way into a site, a building, or an area within a building.

**invacuation** The special form of evacuation that seeks to make use of internal refuges.

**invocation** A formal notification that a contingency plan is to be adopted.

**maximum tolerable missing persons (MTMP)** A target in relation to accounting for everyone in an emergency situation.

**maximum tolerable period of exposure (MTPE)** The absolute maximum time which can be allowed to elapse before everybody has reached safety.

**muster point** A particular spot within an emergency assembly area to which personnel report their presence or in which they gather for further information or instructions.

**personal emergency evacuation plan (PEEP)** A personal emergency or egress plan, normally customized and therefore specific to the situation and needs of a particular individual.

**physical risk assessment (PRA)** The process of identifying threats which could trigger the need for an emergency evacuation.

**primary threat** A trigger event or situation that may be the direct cause of an emergency.

**recovery plan** A plan designed to support the resumption of a specific essential operation, function, or process of an enterprise. Traditionally referred to as a disaster recovery plan (DRP).

**recovery site** A designated site for the recovery of critical functions or operations.

**recovery team** A team of people, assembled in an emergency, which is charged with recovering an aspect of an enterprise, or obtaining the resources required for the recovery.

**resilience** The ability of an organization or a system to anticipate, absorb, respond to, and even profit from the impact of risks and changes. In a BC or emergency context, resilience would include enhancing the effectiveness of the organization through improved availability, rapid response to potential disruptions, and continuity of operations with limited negative effects.

**risk** The possibility that an event or action may result in an undesirable outcome such as loss, damage, or injury.

**risk assessment** A process that is used to identify potential risks to an enterprise.

**risk management** An ongoing set of processes which seek to identify potential causes of risks to an enterprise, understand the possible effects, and reduce or contain the threats to the enterprise.

**risk reduction** The implementation of the preventative measures to address known risks.

**safe evacuation distance (SED)** How far people must move, or be moved, to ensure their safety.

**salvage** The long-term recovery of the equipment, systems, and infrastructure of a site that has been subject to a disaster.

**secondary threat** An event or situation that may occur as a result of an emergency, which would cause further damage or harm.

**site access denial** Any disturbance or activity within the area surrounding a site which restricts access or renders the site unavailable.

**stand down** Formal notification that the alert may be called off because the state of emergency or the disaster situation is over.

**terms of reference (TOR)** A formal description of the purpose and structure of a project, program, or committee.

**test** A procedure which determines whether something works, or is fit for purpose and produces a specific answer.

# Index

## B

## D

## E

# F

## G



## H



## I

## T



## U

## V



## W



## Z

# Credits

**Kristen Noakes-Fry, ABCI,** is Editorial Director at Rothstein Associates Inc. Previously, she was a Research Director, Information Security and Risk Group, for Gartner, Inc.; Associate Editor at Datapro (McGraw-Hill); and Associate Professor of English at Atlantic Cape College in New Jersey. She holds an M.A. from New York University and a B.A. from Russell Sage College.


| | |
|---|---|
| **Cover Design and Graphics:** | Sheila Kwiatek, Flower Grafix |
| **Page Design and Typography:** | Jean King |
| **Index:** | Enid Zafran, Indexing Partners, LLC |

| | |
|---|---|
| **Title Font:** | Nueva STD |
| **Body Fonts:** | Sabon and Frutiger |

# About the Author

Jim Burtles, KLJ, MMLJ, FBCI, is a well-known figure in the Business Continuity profession with over 35 years of experience spread across 24 countries. He was one of founders of the Business Continuity Institute (BCI), where as an honorary fellow he now serves on the Global Membership Council representing the interests of the worldwide membership.

He received the freedom of the City of London in 1992 and was presented with a Lifetime Achievement Award by his peers in 2001. In 2005, he was granted the rank of a Knight of Grace in the Military and Hospitaller Order of St Lazarus of Jerusalem, an ancient charitable body concerned with the treatment of skin diseases.

His first involvement with disaster recovery was in 1974 as a field engineer responsible for repairing and recovering a critical banking system struck by lightning. He went on to become IBM's disaster recovery country specialist before joining a disaster recovery service as their principal emergency management consultant in 1987.

In 2001 he set out on his own as the principal of Total Continuity Management, where he now focuses on executive level training and the development of specialist emergency response skills.

Throughout all these years he tended to specialize in serving the personal needs of those involved in emergency situations. He trained and served as a trauma counselor before developing a counseling technique aimed at helping the victims of disaster-related trauma. This was published in 1998 as A Counselor's Guide to the Restabilization Process. His practical experience includes hands-on recovery work with victims of such violent events as riots, bombings, earthquakes, storms, and fires. This includes technical assistance and support in 90-odd disasters, as well as advice and guidance for clients in over 200 emergency situations.

Through his activities as a trainer and consultant, he has helped to introduce the personal aspects of business continuity, emergency planning, and related disciplines into both the public and private sectors. He is the author of Principles and Practice of Business Continuity: Tools and Techniques (Rothstein Associates, 2007).

Jim Burtles can be contacted at: j.burtles@hotmail.com
or +44 (0)207 289 4491.

# How to Get Your FREE DOWNLOAD of Bonus Resource Materials for This Book

***You are entitled to a free download of the EEP Toolkit*** that accompanies your purchase of Emergency Evacuation Planning for Your Workplace: From Chaos to Lifesaving Solutions, by Jim Burtles.

The EEP Toolkit Download ***includes models and templates in editable formats*** for emergency evacuation planning. It also contains sample plans, reports, questionnaires, and self-assessment questions... and more!

To access these materials is easy — just login to our website as an existing user or register as a new user, and then register your book by following these simple instructions.

*IT'S EASY –*
*LOGIN OR REGISTER YOURSELF ON OUR WEBSITE*

1. FIRST, login as an existing user or register as a new user at www.rothstein.com/register. New users will receive an email link to confirm.

*THEN REGISTER YOUR BOOK*

2. Logging in or registering takes you to our Product Registration page. You'll see a list of books. Simply select your book by clicking the corresponding link to the left and just follow the instructions. You will need to have this book handy to answer the questions.
3. You will receive a confirming email within a few hours with additional information and download instructions.
4. Your registration will also confirm your eligibility for future updates and upgrades if applicable.

***If you have any questions or concerns, please email or call us:***

**Rothstein Associates Inc., Publisher**
203.740.7444 or 1-888-ROTHSTEin fax 203.740.7401
4 Arapaho Road Brookfield, Connecticut 06804-3104 USA
Email: info@rothstein.com
www.rothstein.com